A CULTURAL HISTORY OF CHEMISTRY

VOLUME 4

A Cultural History of Chemistry
General Editors: Peter J.T. Morris and Alan J. Rocke

Volume 1
A Cultural History of Chemistry in Antiquity
Edited by Marco Beretta

Volume 2
A Cultural History of Chemistry in the Middle Ages
Edited by Charles Burnett and Sébastien Moureau

Volume 3
A Cultural History of Chemistry in the Early Modern Age
Edited by Bruce T. Moran

Volume 4
A Cultural History of Chemistry in the Eighteenth Century
Edited by Matthew Daniel Eddy and Ursula Klein

Volume 5
A Cultural History of Chemistry in the Nineteenth Century
Edited by Peter J. Ramberg

Volume 6
A Cultural History of Chemistry in the Modern Age
Edited by Peter J.T. Morris

A CULTURAL HISTORY OF CHEMISTRY

IN THE EIGHTEENTH CENTURY

VOLUME 4

Edited by
Matthew Daniel Eddy and Ursula Klein

BLOOMSBURY ACADEMIC
LONDON • NEW YORK • OXFORD • NEW DELHI • SYDNEY

BLOOMSBURY ACADEMIC
Bloomsbury Publishing Plc
50 Bedford Square, London, WC1B 3DP, UK
1385 Broadway, New York, NY 10018, USA
29 Earlsfort Terrace, Dublin 2, Ireland

BLOOMSBURY, BLOOMSBURY ACADEMIC and the Diana logo are trademarks of
Bloomsbury Publishing Plc

First published in Great Britain 2021
Paperback edition published in 2025

Copyright © Bloomsbury Publishing Plc, 2025

Cover design: Rebecca Heselton
Cover image © DeAgostini/Getty Images

All rights reserved. No part of this publication may be reproduced or transmitted in any form or by any means, electronic or mechanical, including photocopying, recording, or any information storage or retrieval system, without prior permission in writing from the publishers.

Bloomsbury Publishing Plc does not have any control over, or responsibility for, any third-party websites referred to or in this book. All internet addresses given in this book were correct at the time of going to press. The author and publisher regret any inconvenience caused if addresses have changed or sites have ceased to exist, but can accept no responsibility for any such changes.

A catalogue record for this book is available from the British Library.

A catalog record for this book is available from the Library of Congress.

ISBN: PB: 978-1-3505-5214-2
       Pack: 978-1-3505-5229-6
       ePUB: 978-1-3502-5153-3
       ePDF: 978-1-3502-5152-6

Series: The Cultural Histories Series

Typeset by Integra Software Services Pvt. Ltd.
Printed and bound in Great Britain

To find out more about our authors and books visit www.bloomsbury.com and sign up for our newsletters.

CONTENTS

List of Illustrations   vii

Series Preface   x

Introduction: The Core Concepts and Cultural Context of
Eighteenth-Century Chemistry   1
Ursula Klein and Matthew Daniel Eddy

1 Theory and Concepts: Transformations of Chemical Ideas
in the Eighteenth Century   23
Ursula Klein

2 Practice and Experiment: Operations, Skills, and Experience
in Eighteenth-Century Chemistry   45
Victor D. Boantza

3 Laboratories and Technology   71
Marco Beretta

4 Culture and Science: Chemistry in its Golden Age   93
Bernadette Bensaude-Vincent

5 Society and Environment: Chemistry and Daily Life during
the Eighteenth Century   113
Matthew Daniel Eddy

6 Trade and Industry: An Era of New Chemical Industries
and Technologies   137
Leslie Tomory

7   Learning and Institutions: Didactic Chemistry and
    Practical Instruction   157
    John C. Powers

8   Art and Representation: Cultural Modalities of Chemistry
    in the Eighteenth Century   175
    John R.R. Christie

NOTES   203
BIBLIOGRAPHY   206
LIST OF CONTRIBUTORS   229
INDEX   230

LIST OF ILLUSTRATIONS

2.1 Various distillation retorts, pelicans, alembics, furnaces, and other chemical apparatus. From Nicolas Lémery, *Cours de chymie, contenant la manière de faire les opérations qui sont en usage dans la médecine ... Nouvelle édition, revue, corrigée & augmentée* (Paris: d'Houry, 1757), plates 1, 2, 5, and 6. Courtesy of HathiTrust   51

2.2 Geoffroy's 1718 affinity table. From E.F. Geoffroy, "Table des differents rapports observés entre differentes substances," *Mémoires de l'Académie Royale des Sciences* (Paris: Imprimerie Royale, 1718), p. 212, plate 8. Sourced from Wikipedia   59

2.3 Hales' two-vessel apparatus for measuring amounts of gas produced or absorbed; pedestal apparatus; iconic pneumatic trough. From S. Hales, *Statical Essays: Containing Vegetable Staticks; Or, An Account of some Statical Experiments ... Also, a Specimen of An Attempt to Analyse the Air, by a great Variety of Chymio-Statical Experiments*, 3rd ed. (London: Innys and Manby, 1738), vol. 1, pp. 168, 211, and 266. Courtesy of HathiTrust   62

2.4 Cavendish's pneumatic instruments and method. From H. Cavendish, "Three Papers, Containing Experiments on Factitious Air," *Philosophical Transactions of the Royal Society* 56 (1766), table VII, p. 141. Courtesy of Royal Society Publishing   64

2.5 Priestley's pneumatic instruments and workshop. From J. Priestley, *Experiments and Observations on Different Kinds of Air, and other branches of natural philosophy ... Being the former six volumes abridged and methodized, with many additions* (Birmingham: Pearson and Johnson, 1790), vol. 1, plates I and II. Courtesy of HathiTrust ... 65

2.6 Examples of Lavoisier's instruments from the early 1770s (left) and late 1780s (right), including his ice calorimeter (bottom right) for quantifying heat in chemical reactions. A.-L. Lavoisier, *Opuscules physiques et chymiques* (Paris: Durand, 1774), vol. 1, plates I and II; A.-L. Lavoisier, *Elements of Chemistry, in a New Systematic Order, Containing All the Modern Discoveries*, trans. R. Kerr (Edinburgh: Creech, 1790), plate XI. Courtesy of HathiTrust ... 68

3.1 A chemical laboratory from an engraving in the *Encyclopédie*, 1763. Courtesy the ARTFL Encyclopédie Project, University of Chicago ... 76

3.2 A.-L. Lavoisier's gasometer, from an engraving made by Marie Lavoisier and published in the second volume of A.-L. Lavoisier, *Traité élémentaire de chimie*, 1789. Author's copy ... 89

3.3 A.-L. Lavoisier in his laboratory conducting an experiment on the respiration of a resting man. Photogravure (ca. 1850) after a drawing (ca. 1790) by Marie Lavoisier. Courtesy Wellcome Collection CC BY ... 90

5.1 William Hogarth, *In the Cabinet of the Quack doctor, the Viscount Squanderfield Holds Out a Small Pill-Box as a Young Girl Dabs Her Face with a Handkerchief*, colored aquatint after William Hogarth. Published in France and based on William Hogarth's original (London: 1745). The print was Plate III in Hogarth's series entitled *Marriage-a-la-Mode*. Courtesy Wellcome Collection CC BY ... 115

5.2 A female figure performing chemical experiments with a furnace: representing chemistry. Etching by E.-J.-N. de Ghendt after C.-N. Cochin the younger, 1773. Courtesy Wellcome Collection CC BY ... 119

5.3 Larévellière-Lépeaux sits in a disordered quack doctor's room, in the presence of seven wounded French generals, one of them vomiting; representing French defeats in 1799 and Bonaparte's failed imperial ambitions in the east. Colored etching by J. Gillray, 1799. Courtesy Wellcome Collection CC BY ... 126

5.4 Mount Vesuvius emitting a column of smoke after its eruption on August 8, 1779. Colored etching by Pietro Fabris, 1779. Courtesy Wellcome Collection CC BY ... 130

LIST OF ILLUSTRATIONS

5.5 Four men sit round the tax man and blow smoke in his face. Colored aquatint, late eighteenth century, after G. M. Woodward (?). Courtesy Wellcome Collection CC BY — 134

6.1 Gobelins Dye Works, from *Encyclopédie, ou dictionnaire raisonné des sciences, des arts et des métiers*, eds Denis Diderot and Jean le Rond d'Alembert (1772), vol. 10, plate VIII, colored tint from ca. 1800. Sourced from Wikimedia Commons, Public Domain — 142

6.2 Alum works, from *Encyclopédie, ou dictionnaire raisonné des sciences, des arts et des métiers*, eds Denis Diderot and Jean le Rond d'Alembert (1768), plate, vol. 5, p. 26. Sourced from Wikimedia Commons, Public Domain — 144

8.1 Joseph Black lecturing at Glasgow. Etching by John Kay, 1787. Photograph by Universal History Archive/Getty Images — 177

8.2 Laboratory of William Lewis, *Commercium Philosophico-Technicum*, 1763. Sourced from Wikimedia Commons, Public Domain — 182

8.3 Laboratory of Conrad Barchusen, der Sonnenborgh, Utrecht, from Barchusen, *Elementa chemiae* (Lugduni Batavorum, 1718). Courtesy of Science History Institute — 183

8.4 *The Alchemist in Search of the Philosophers Stone*, Joseph Wright of Derby, 1771. Derby Museum. Photograph by Christophel Fine Art/Universal Images Group via Getty Images — 185

8.5 Portrait of the French chemist Antoine Laurent de Lavoisier with his wife (Marie-Anne-Pierrette Paulze). Painting by Jacques Louis David, 1788. The Metropolitan Museum of Art, New York. Photograph by Leemage/Corbis via Getty Images — 187

8.6 Drawing by Marie Lavoisier of an experiment on respiration, Salpetrière laboratory, 1790–1791. Science History Images/Alamy Stock Photo — 189

8.7 James Sayers, "A Vision. The Repeal of the Test Act," 1790. Courtesy Alamy Stock Photo — 192

SERIES PREFACE

A Cultural History of Chemistry examines the history of chemistry and its wider contexts from antiquity to the present. The series consists of six chronologically defined volumes, each volume comprising nine essays; these fifty-four contributions were written and/or edited by a total of fifty scholars, of ten different nationalities. Of Bloomsbury's many six-volume *Cultural Histories* currently in print, this is the first in the physical or natural sciences; it is also the first multivolume history of chemistry to appear since James Riddick Partington's four-volume *History of Chemistry* concluded more than fifty years ago. It is distinguished, among other qualities, by its endeavor to take the subject from antiquity right to the present day.

This is not a conventional history of chemistry, but a first attempt at creating a cultural history of the science. All cultures, including the various branches of natural science, consist of mixed constructs of social, intellectual, and material elements; however, the cultural-historical study of chemistry is still in an early stage of development. We hope that the accounts presented in these volumes will prove useful for students and scholars interested in the subject, and a starting point for those who are striving to create a more fully developed cultural history of chemistry.

Each volume has the same structure: starting with an interpretive overview by the volume editor(s), the eight succeeding chapters explore for each respective era in chemistry its theory and concepts; practice and experiment; laboratories and technology; culture and science; society and environment; trade and industry; learning and institutions; and art and representation. Readers therefore have the option to read multiple chapters in a single volume, thus learning about the cultures of chemistry in a single era; or they may prefer instead to read corresponding chapters across multiple volumes, learning about (e.g.) the art and representations of chemistry through the ages. Though the scope is global, major emphasis is placed on the Western tradition of science and its contexts.

Whether read synchronically or diachronically, in any multiauthor undertaking like this one readers will inevitably notice overlaps and repetitions, conflicting historical interpretations, and (despite the magnitude of the project) occasional gaps in coverage. These are inescapable consequences, but they actually offer advantages to the reader, both in making each chapter closer to self-contained and in demonstrating the dynamism of the discipline; like science itself, the study of its history is ever contested and incomplete.

Chemistry has been called the "central science," due to its fundamental importance to all the other physical and natural sciences. It is the archetypical science of materials and material productivity, and as such it has always been deeply embedded in human industry, society, arts, and culture, as these volumes richly attest. The editors and authors hope that *A Cultural History of Chemistry* will be of great interest and enjoyment not just to chemists and specialist historians of science, but also to social, economic, intellectual, and cultural historians, as well as to other interested readers.

Peter J.T. Morris and Alan J. Rocke
London (UK) and Cleveland (USA)

Introduction

The Core Concepts and Cultural Context of Eighteenth-Century Chemistry

URSULA KLEIN AND MATTHEW DANIEL EDDY

Before the late seventeenth century, chemistry was neither a professional occupation nor an academic discipline. Persons recognized as alchemists or chemists (collectively: "chymists") were individuals scattered over different places and professions. Making a good living as a chymist was achieved mainly at princely courts. Add to this the fact that the medieval and Renaissance university did not include chymistry in its curriculum. After the introduction of medical chymistry in the late sixteenth and seventeenth centuries, several universities included courses on chymistry in their medical curricula. This made it possible for professors of medicine, physicians, and apothecaries to earn money by preparing chemical remedies. But how did chymistry expand from being a philosophy and an art practiced in a variety of local contexts into a fully fledged chemical discipline? This introductory chapter answers this question by presenting an overview of the core concepts and cultural context that enabled chemists to transform chemistry into an independent discipline that systematically collected local knowledge about a broad range of material substances and ways of their change, and how it further engendered new empirical and theoretical knowledge.

INSTITUTIONALIZATION OF CHEMISTRY AND THE FORMATION OF CHEMICAL COMMUNITIES

Before the late seventeenth century, the book, in printed or manuscript form, was a central mode of communication through which chemical knowledge was accumulated, structured, and socially transmitted. Many historians see Andreas Libavius' *Alchemia*, published in 1597, as the first chemical textbook that presented the terrain of chemical knowledge in a comprehensive and systematic way (Hannaway 1975). It stands at the beginning of a long tradition of chemical textbooks, especially in France, that contributed to the formation of the discipline of chemistry, in the sense of a distinctive body of knowledge that assembles information from many different sites and is demarcated from other institutionalized systems of knowledge.

In the eighteenth century, the institutionalization of chemistry as an independent discipline took off. Chemists entered scientific academies in many European countries, and the number of university professors of chemistry quickly grew, both in Europe and its colonies. At the Jardin Royal des Plantes in Paris, two chairs in chemistry were founded in the seventeenth century and sustained a teaching tradition in the eighteenth century. In the artillery schools, mining academies, schools of civil engineering, agricultural schools, medical colleges, and chemical–pharmaceutical schools founded in the second half of the eighteenth century in continental Europe, chemistry occupied an even more important place (Klein 2020). At the same time the book market for chemical literature expanded, and public lectures of chemistry became fashionable. The Enlightenment movement greatly fostered the curiosity for chemistry (see below). While before the eighteenth century only the intellectual elite, mostly men, and well-to-do people read chemical texts, books and distinct journals about chemistry now flooded the market and were also read by women. In this way, knowledge about unfamiliar materials, chemical operations, and instruments as well as chemical concepts and theories were transmitted across different sites and social strata.

Eighteenth-century chemical textbooks circumscribe quite well the body of trans-local knowledge accumulated and structured within the institutionalized discipline of chemistry. These textbooks differed, however, from today's globalized chemical textbooks, as they had also some national characteristics (Abbri and Bensaude-Vincent 1995). In France, they often started with a brief chapter on chemical "theory" or chemical "philosophy." Early eighteenth-century chemists wanted to know, in particular, whether nature created the plurality of different substances out of a few basic constituents, or chemical "principles," which caused the substances' properties. In the second half of the eighteenth century, this theoretical agenda was supplemented by theories of chemical change, ranging from ideas about law-like relationships, or chemical "affinities," to explanations of major types of chemical reaction (see below).

All of the eighteenth-century French chemical textbooks also included a chapter on experimental techniques and chemical instruments. However, most French chemical textbooks, and more or less all chemical textbooks at the time, were concerned with something else: the description, or "history," of chemical substances. Eighteenth-century chemists filled hundreds of pages of their textbooks with descriptions of substances, their properties, techniques of their preparation, methods of their unambiguous identification and classification, their practical uses, and their chemical reactions. For example, the twelfth French edition of Nicolas Lémery's popular *Cours de Chymie*, published in 1724, devotes nine pages to theoretical principles, some thirty pages to chemical furnaces, vessels, and techniques (including many pictures), and almost 700 pages to substances. Like the early modern natural-historical museums, eighteenth-century chemical textbooks were primarily collections of empirical knowledge about material substances. Their main goal was to present in a comprehensive way all relevant knowledge about material substances, the methods of their preparation, their recurrent properties, uses, and manifold reactions.

The institutionalization of chemistry in eighteenth-century Europe went hand in hand with the formation of distinct communities of chemists. In the sixteenth and seventeenth centuries, chemical books could be written by authors who lived in isolation from like-minded people. By contrast, a chemical community was emerging around 1700 which began to create the working life of a discipline. Karl Hufbauer produced an exemplary study of the formation of the German chemical community between 1720 and 1795. He showed that the publication of chemical journals in the last third of the eighteenth century was a crucial step in the consolidation of the German chemical community (Hufbauer 1982). Lorenz Crell's *Chemische Annalen*, published from 1778, was the immediate inspiration for the French *Annales de chimie*, launched in 1789 (Crosland 1994). The eighteenth-century chemical journal became a forum for collecting, discussing, and distributing new chemical knowledge, and it helped to define the boundaries of the chemical discipline. As Hufbauer has also observed, recognition of a person as a chemist depended on criteria that were quietly fixed in the emerging chemical communities. In the eighteenth century, many chemists came from the pharmaceutical and medical professions, which implied that they possessed a laboratory and practical knowledge of chemistry. The teaching of chemistry was another important criterion for recognizing a chemist, which demonstrated that the person had mastered not only a small segment of practical chemistry but rather the entire discipline. Possession of a laboratory, publication of experimental chemical reports, and citations of publications by other chemists were additional criteria. Furthermore, social status as well as personal communication and correspondence with other chemists also contributed to social recognition as a chemist (Hufbauer 1982;

Klein 2007, 2010). As the chemical communities in the different European countries were in contact with each other, it is no exaggeration to speak of a community of European chemists in the eighteenth century.

The institutionalization of chemistry as an academic discipline and the formation of chemical communities in the eighteenth century also stimulated reflections on the question of what constituted the distinctiveness of the chemical discipline vis-à-vis neighboring disciplines, particularly medically related fields such as materia medica and botany, but also mineralogy and the emerging discipline of physics, and how chemistry related to the more comprehensive field of experimental natural philosophy. Learned discourse on these issues was often concerned with the applicability of the mechanical–corpuscular philosophy to chemistry. For example, a few years after the foundation of the Paris Academy of Science, at the end of the 1660s, the Academy's chief chemist, Samuel Cottereau Duclos, discussed the difference between chemistry and the areas of natural philosophy described by mechanical concepts and laws. His demarcation of the chemical discipline from mechanics and physics implied a critique of Robert Boyle's application of the mechanical–corpuscular philosophy to chemistry. His critique was echoed by the influential German chemist Georg Ernst Stahl and many other eighteenth-century chemists (Boantza 2013b).

PRACTICAL CHEMISTRY

Alongside the institutionalization of chemistry at scientific academies, universities, and technological colleges, practical chemical knowledge continued to thrive, firmly embedded as it was in local artisanal crafts and in the emerging chemical industry, and it remained an important source of the institutionalized chemical discipline. Metal production and metal working as well as the preparation of so-called chemical remedies by means of distillation, extractions, and other kinds of chemical operations were the most important practical pillars of institutionalized chemistry. As these crafts engendered not only local chemical knowledge but also instruments, techniques, and material substances, and even shared the laboratory with the institutionalized chemical discipline, we might designate these artisanal activities as chemical arts or "practical chemistry." Practical chemistry extended to household production and to some additional arts and crafts, most notably brewing and wine making, dyeing and calico printing, the making of glass, ceramics, and porcelain, the making of gunpowder, and the preparation of a plethora of new luxury goods such as artificial mineral waters, cosmetics, liqueurs, artificial gems, and fashionable drugs such as ether. Hence, in contrast to the chemical discipline promoted in specific institutions, practical chemistry was always local knowledge produced at many different sites (Fors 2003; Klein and Spary 2010; Roberts and Werrett 2017).

In the framework of the eighteenth-century discipline of chemistry, anchored at scientific academies, universities, and technical schools, practical chemical knowledge stemming from local artisanal sites was integrated into a more comprehensive body of disciplinary knowledge. It was thereby also transformed into more general knowledge. Elements of practical chemical knowledge regarded as peculiar to local circumstances were eliminated, while knowledge deemed to be relevant across different locations was preserved (Klein 2020). This active abstraction from accidental local features of practical chemical knowledge concerned especially material substances. Unlike today, the substances eighteenth-century chemists described in their textbooks and experimented with in their laboratories were not standardized, pure chemical substances. In local workshops and everday life chemists nearly always used some admixtures of other kinds of substances, which were, however, relevant for practical applications. By contrast, when chemists regarded iron as a "chemical" substance, they abstracted not solely from numerous properties of the local varieties of iron, they also focused on those properties that were observable in iron's behavior in fire and when it was mixed with other substances. For the same reason, early-eighteenth-century chemists believed, for example, that the commercial varieties of potash and natron (soda) were irrelevant in an academic context and hence that there was just one chemical substance that they named "fixed alkali." Abstraction and generalization of practical knowledge about materials allowed chemists to focus their attention on the recurrent properties of chemical substances and on regularities in their chemical reactions, which paved the way for the construction of affinity tables and chemical theory.

Chemical knowledge also traveled in the other direction, that is, from the institutionalized chemical discipline to local artisanal and industrial sites. The eighteenth century saw the emergence of particular sub-disciplines of chemistry that were tailored to specific practical demands, such as metallurgical chemistry, thriving in the context of mining academies; pharmaceutical chemistry, taught at chemical–pharmaceutical boarding schools; and agricultural chemistry, promoted not only in the context of the patriotic and economic societies but also at some universities (Schleiff and Konečný 2013; Klein 2015b; Klein 2020). Hence, eighteenth-century chemists held that they produced useful knowledge. By the end of the century, there were distinct textbooks on "technological chemistry." Chemists also introduced the distinction between "pure" and "applied chemistry." Pure chemistry covered the core areas of the chemical discipline that were taught to all students of chemistry. Applied chemistry comprised the ever-extending field of more specialized chemical–technological knowledge, which corresponded with an extending audience of chemistry (Meinel 1983). The eighteenth-century chemical–pharmaceutical schools, mining academies, and other technical colleges educated and trained technical experts, such as apothecaries, mining officials, and dyers. They thereby contributed to a slow transformation of practical chemistry and the emergence of the modern chemical industry (Klein 2020).

STUDIES OF SUBSTANCES

Eighteenth-century chemistry was primarily a science of material substances and their reactions. In their laboratories, chemists decomposed substances, analyzed their components, re-synthesized the initial compound, and synthesized new substances. Experiments with material substance came in many different guises and were by no means absolutely timeless activities. For example, alchemists and eighteenth-century chemists had different conceptions of the chemical substances made by fire and extracted from natural raw materials by means of solvents. While alchemists understood such chemically prepared substances as "subtle" spirits and essences that were deprived of the "corporality" of the original substances and possessed enhanced properties, eighteenth-century chemists dismissed ontological distinctions between enhanced, "subtle" substances and more "corporeal" ones.

Many of the substances that early-eighteenth-century chemists described in their textbooks and used in their laboratory experiments had long been produced in the arts and crafts (Multhauf 1966). Gold, silver, copper, lead, iron, tin, and mercury were metals of commerce, smelted from ores since antiquity. Likewise, for millennia the metal alloys bronze (a copper–tin alloy), brass (a copper–zinc alloy), and electrum (a naturally occurring gold–silver alloy) had been used for the making of weapons, tools, coins, and jewelry. Bismuth and antimony were regularly produced in the sixteenth century to be used as ingredients for the preparation of Paracelsian remedies, and arsenic and zinc soon followed them. Not all of these substances were originally defined as "metals," not least because of the ancient dogma that associated seven metals with the sun, moon, and the five then-known planets (Venus, Mars, Saturn, Mercury, and Jupiter). In his *Elements of Chemistry* (1789), Antoine-Laurent Lavoisier listed seventeen metals, among which cobalt, manganese, molybdenum, nickel, platinum, and tungsten had been discovered in the eighteenth century.

Apart from metals, eighteenth-century chemists described and experimented with dozens of additional materials that strike us today as typical "chemical substances." To these chemical substances belonged natron, an impure soda (sodium carbonate) that came from Egyptian mines or the leaching of marine plant ashes (then named "barilla"), as well as potash (potassium carbonate), produced from inland plants. These two "alkalis," which early eighteenth-century chemists did not distinguish as different chemical substances, were used as ingredients for the manufacture of soap and glass, as chemical agents in the cleansing and dyeing of textiles, and as ingredients for the preparation of chemical remedies. Early-eighteenth-century chemists also experimented with different kinds of "earths" such as limestone, or "calcareous earth," and "vitrifiable earth," which were employed in the manufacture of glass, ceramics, and artificial gems. Quicklime, made by heating limestone, was already produced in antiquity on a larger scale, as it was the principal ingredient of mortar for building construction; it was also used in the production of glass and leather. Another substance related to limestone was gypsum, used in construction projects since antiquity. In the middle of the eighteenth century, the Prussian chemist Johann Heinrich Pott recognized four different kinds of "earth," namely, *terra alcalina* (limestone), *terra gypsea* (gypsum), *terra argillacea* (clay), and *terra vitrescibilis* (silicia), based on numerous analyses he had carried out on behalf of King Frederick II ("the Great"), who had commissioned him to discover the secret of making porcelain.

Among eighteenth-century chemists' "salts" – such as common salt, vitriol, alum, saltpetre, sal ammoniac, and borax – many were long known in households and the arts and crafts. Common salt, vitriol, and alum had been commercial goods since antiquity. Vitriol was used for its property, with oak galls, of giving a black color to inks and leather, and alum helped in fixing vegetable dyes to cloth. From the thirteenth century, saltpetre was an important ingredient, with sulfur and charcoal, for the making of gunpowder and for the preparation of aqua fortis (nitric acid). Sal ammonic (ammonium chloride) and borax (sodium tetraborate) were traded by the Venetians and applied in the refining of precious metals, dyeing and leather tanning (sal ammoniac), and in enamel glazing (borax). All of these salts were also applied pharmaceutically, and from the sixteenth century many of them were used for preparing new mineral acids, which chemists also classified as "salts." In the late eighteenth century, vitriolic acid (sulfuric acid) and other mineral acids were produced on a larger scale and promoted the emergence of the chemical industry (Clow and Clow 1952; Smith 1979).

PLANT AND ANIMAL CHEMISTRY, MINERALOGICAL CHEMISTRY, AND STUDIES OF GASES

Metals, alkalis, earths, salts, and mineral acids were not immediately found in nature, but were products of labor, processed from natural raw materials in mines, foundries, workshops, and laboratories. The alchemical laboratories yielded numerous additional chemical substances – among them processed vegetable and animal substances such as balsams, resins, gums, waxes, essential oils, fats, and alcoholic spirits – that were also applied as chemical remedies. Well into the eighteenth century, the drug lists of physicians and apothecaries provided chemists with a huge repertoire of substances on which to work. This applied not only to the processed "chemical substances," but also to the natural raw materials that eighteenth-century chemists also analyzed in their laboratories.

In addition to their studies of processed chemical substances, eighteenth-century chemists analyzed a plethora of natural raw materials, ranging from entire plants, roots, leaves, flowers, and vegetable seeds, to bones, hoofs and horns, fat, blood, milk, and other raw materials of animal origin. Many of these natural materials were exotic goods imported from colonies. The chemical analysis of milk stemming from humans, cows, goats, and sheep, as well as donkeys and camels, were among the most prominent chemical–medical enterprises in eighteenth-century Europe (Orland 2010). In the second half of the eighteenth century, chemists often demarcated these studies as "plant and animal chemistry." At the end of the eighteenth century, plant and animal chemistry was transformed into modern organic chemistry (Holmes 1989; Klein and Lefèvre 2007).

Eighteenth-century chemists also analyzed mineral waters and natural raw minerals such as stones (Eddy 2008; Eddy, 2010). In the second half of the eighteenth century, the chemical analysis of raw minerals contributed to the formation of a new chemical sub-discipline: mineralogical chemistry or chemical mineralogy, which promoted the discovery of numerous new chemical substances. The Swedish chemist and mineralogist Axel Frederick Cronstedt has often been called the founder of chemical mineralogy. His systematic use of the blowpipe for mineral analysis and his discovery of nickel in 1751 made him well known in the learned world (Abney Salomon 2019). Cronstedt discovered nickel through the analysis of ores he had obtained from the cobalt mines at Los, in Hälsingland. At the time, he was a leading mining expert of the Bureau of Mines (*Bergskollegium*), the governmental agency that guided and controlled the Swedish mining industry and metal production, and he also carried out his analyses in the laboratory of the silver works at the Skiss foundries in central Sweden. Torbern Bergman, the most famous Swedish analytical chemist of the eighteenth century, also analyzed numerous raw minerals. He was professor of chemistry at the University of Uppsala, and he had also connections to the Bureau of Mines and mining industry (Szabadváry 1966; Fors 2003). The mining industry was also an important context of chemical mineralogy outside of Sweden, for example, in Austria–Hungary and in Saxony.

In France the analysis of minerals preoccupied chemists from the 1790s, after the École des Mines had been founded. René-Just Haüy, who was a founder of chemical crystallography and revised the system of mineralogical classification, taught physics and mineralogy at the École des Mines from 1795 (Mauskopf 1976). The most skilled chemical analyst in late-eighteenth-century France, Nicolas Louis Vauquelin, analyzed numerous natural minerals and discovered two new substances in 1798, chromium and beryllia. He was also a professor of assaying at the École des Mines, an official assayer of precious metals for Paris, and an inspector of mines. Haüy and Vauquelin exerted a strong influence on Antoine-François de Fourcroy, collaborator of Lavoisier, who presented in his eleven-volume textbook *A General System of Chemical Knowledge* (1804) a long section on chemical petrography, including tables of recently discovered quantitative compositions of stones.

In the German states, the apothecary-chemist Martin Heinrich Klaproth became the most famous chemical analyst of raw minerals after he had entered the circle of the Minister Friedrich Anton von Heinitz, who headed the Prussian mining administration. In 1789, Klaproth discovered uranium and zirconia in ores that he had obtained from the Mining Academy of Freiberg in Saxony. This was followed by the discovery of the "earths" (later: metal oxides) of strontium, chromium, and cerium, and by contributions to the discovery of tellurium and the earths of beryllium and titanium. Even in the last years of his life, he conducted innumerable quantitative chemical analyses and determined the chemical composition of over 200 substances, most of which were raw minerals (Klein 2014b).

In the course of the eighteenth century chemists established yet another new field of chemical inquiry: the chemistry of "airs" or "gases" (Brock 1993; Levere 2000; Kim 2003; Boantza 2013a). In 1727, Stephen Hales found a method to isolate the "air" produced from a heated solid substance and estimate its amount. He purified the air by passing it through water and then collected it in a suspended vessel by the downward displacement of water. While Hales still believed that there was just one kind of air, Joseph Black soon demonstrated that "fixed air" (carbon dioxide) was different from ordinary air. In the 1770s, Henry Cavendish, Joseph Priestley, and Carl Wilhelm Scheele improved Hales' apparatus to study different varieties of air. They discovered vital air (oxygen), phlogisticated air (nitrogen), and inflammable air (hydrogen). These discoveries were a sensation, because before the work of Black chemists had long believed that there was just one kind of air. Knowledge about different varieties of air and of their chemical reactions contributed significantly to the restructuring of chemistry in the 1770s and 1780s.

THE LABORATORY

The laboratory was the hallmark of chemistry, and experimentalists in the eighteenth century recognized that a chemist needed to own, or have access to, one (Morris 2015). At the time, chemistry was usually the science that routinely carried out experiments in laboratories. Hence, the term "laboratory" referred almost exclusively to a room or building where chemical operations were performed (Klein 2008). A "laboratory" was, however, not exclusively a scientific–chemical institution but also a site of practical chemistry. The Latin word "laborare," from which "laboratory" is derived, meant any kind of manual work, including commercial labor. Accordingly, Diderot and d'Alembert's *Encyclopédie* equated "laboratory" (*laboratoire*) with "shop" (*boutique*), and defined it as a "closed and covered place" that "contains chemical equipment" such as furnaces, vessels, and instruments (quoted in Klein 2008: 771). Similarly, the economic–technical encyclopedia compiled by the German cameralist Johann Georg Krünitz points out that a "laboratory" is a "labour- or workhouse," and that the term is mostly used to designate "the place in chemistry and in pharmacy that is suited for chemical work" and "the house in which gunsmiths and sergeant-artificers manufacture their materials" (quoted in Klein 2008: 771). In the eighteenth century, laboratories thus existed in scientific institutions as well as in the artisanal world, especially in apothecary shops, mines, foundries, mints, arsenals, dyeing manufactories, porcelain manufactories, distilleries, and perfumeries.

Moreover, apart from a few exceptions such as Antoine-Laurent Lavoisier's sophisticated and expensive laboratory equipment, there was a strong similarity between laboratories at academic and at artisanal sites. As a rule, eighteenth-century laboratories were dominated by furnaces, chimneys, distillation apparatus, and equipment for metal smelting and assaying, which demonstrates the importance of fire for carrying out chemical operations. There were many types of furnaces, large immovable ones as well as small portable ones. There were those which yielded high temperatures and which were used only for metal smelting, glazing, and enamelling. There were still more furnaces used for long, gentle heating of vegetable and animal substances, and various others for distillation. The most common furnace was the reverberatory furnace, which had a heating chamber with a hemispherical top in which the heat was reflected to create very high temperatures. Fire hazards were a permanent concern of eighteenth-century chemists. Hence, chemists designed laboratories with vaulted ceilings made of stone. Additionally, laboratories were usually located on the ground floor of the house or in a courtyard, which was also convenient for the transport of wood, charcoal, and water.

The basic furnishings of eighteenth-century laboratories were a table, shelves, and a closet for storing vessels, small instruments, and chemicals. Chemists used various kinds of instruments and vessels for carrying out small-scale table-top experiments such as solution in acids and subsequent precipitation. Among these instruments were small beakers, glass jars, matrasses with a flat bottom (Florence flasks), crystallization dishes, stirring rods, and filters. Alembics, pelicans, retorts, and receivers were used for distillation; crucibles and muffles were the most ubiquitous instruments for carrying out metallurgical operations; and bottles and glass phials were used for storing chemicals. Mortars and pestles, spatulas, a water barrel, water baths (*balneo mariae*), bellows, and balances were equally indispensable instruments in any chemical laboratory. Early-eighteenth-century chemical laboratories differed little from their seventeenth-century counterparts. The glass instruments and vessels used from the middle of the century for studying gases was the major technological change of the eighteenth-century chemical laboratory (Eklund 1975; Holmes 1989; Morris 2015).

CHEMICAL THEORY

In the late seventeenth century, the majority of chemists still believed that the great variety of natural substances were generated from simple elements, or "chemical principles," which were few in number (three to five). According to this view, the different kinds of natural substances belonging to the mineral, vegetable, or animal kingdom resulted from the mixing of different proportions of the three, four, or five simple elements and the merging together of their qualities into fully homogeneous natural substances or "mixts." A minority of chemists adopted an atomistic or corpuscular understanding of the origin and nature of natural substances. Their concepts of element, atom, and corpuscle were deeply rooted in philosophical traditions and only loosely linked with artisanal and experimental practice. In the course of the eighteenth century, the relationship between theory and experiment changed, along with the emergence and proliferation of new chemical concepts and theories that systematically interconnected explanations of chemical change with explanations of the formation and composition of substances. The new concepts of chemical combination, compound, composition, and affinity were initially elaborated in a context that was independent of theories about chemical elements and atoms (Klein 1994a; Klein and Lefèvre 2007). This changed only in the late eighteenth century, when Antoine-Laurent Lavoisier introduced what became more or less the modern concept of chemical element and amalgamated it with the compositional view long embodied in affinity tables (Siegfried and Dobbs 1968; Siegfried 2002; Hendry 2019).

Early-eighteenth-century chemists also believed that natural substances, including substances belonging to the mineral realm, were growing and ripening in nature. They regarded numerous cases of chemical change carried out in the laboratory as artificially accelerated imitations of the substances' natural changes, in which fire replaced the Sun and the Earth's natural heat. In accordance with this view, their concepts of transmutation of metals and of extraction of the subtle essences of natural substances presented chemical change as changes of the properties of a single substance. By the middle of the eighteenth century, however, chemists' understanding of chemical change as transformation of properties of just one single substance became an exception rather than the rule. The alternative way of looking at chemical change was interaction, or reaction, between different kinds of substances. In the laboratory, certain substances could be prompted to combine with each other to form new kinds of substances, which could be subsequently decomposed into the original substances. Chemists could also simultaneously decompose two substances and recompose new kinds of substances from their components.

This new understanding of chemical change hinged on operations done in the artisanal world as well as on systematic experiments carried out in academic laboratories, in particular on the chemical preparation of salts and metallurgical operations (Holmes 1988; Klein 1994a; Klein 1994b; Klein and Lefèvre 2007). In the early eighteenth century, the leading chemist of the Paris Academy of Sciences, Wilhelm Homberg, pointed out that acids that had combined with alkalis made up certain kinds of salts, which he called middle salts. These salts, which chemists also prepared from metals or earths, could be decomposed back into their two ingredients. These kinds of chemical preparations had long been known in medical–pharmaceutical practice, but Homberg paid systematic attention to them and explained them in a coherent way. Similarly, smelters and silver- and goldsmiths had long accumulated experience about combinations of different metals into alloys, and the recovery of the original metals in subsequent chemical operations. In 1718, the apothecary-chemist Etienne-François Geoffroy assembled chemists' knowledge about these kinds of reversible reactions in a table of chemical "rapports," or affinities, and explained the results as displacement reactions directed by rapports between tangible chemical substances.

The explanation of chemical change as a process of movement and recombination of tangible substances, in which the substances were preserved in the newly prepared chemical compounds like stable building blocks even though their characteristic properties had disappeared, was to some extent a mechanical understanding of chemical change. However, chemists supplemented this mechanical view by the concept of chemical "rapports" or affinities. Substances could be mechanically mixed in all possible ways, but chemists knew from experience that they did not combine with each other arbitrarily. On the contrary, chemical combination took place only between distinct kinds of substances, and it was even possible that one substance displaced another from its combination and took its place, making a new chemical compound. Hence, chemists assumed that there were different affinities between pairs of different substances that enabled them to combine among each other in a selective way. In order to unravel the regularities, or even laws, of affinity, eighteenth-century chemists arranged displacement reactions into affinity tables (Klein 1994a; Klein 1995; Duncan 1996; Kim 2003; Eddy 2014).

In the middle of the eighteenth century, chemists regarded such chemical affinity tables as their most advanced conceptual tools for explaining chemical change. They also assimilated the phlogiston theory into this framework. The theory, introduced by Georg Ernst Stahl in the early eighteenth century, explained combustion as the separation of the invisible "phlogiston" from the combustible substance. The release of phlogiston into the air, Stahl assumed, was the cause of heat and flames emerging during combustion. In an analogous way, Stahl explained the corrosion, or calcination, of metals as the decomposition of metals into phlogiston and a metal "calx." He also pointed out that it was possible to recover the original metal by heating the metal calx with charcoal, which was rich in phlogiston. Hence, the reduction of a metal calx into its corresponding metal was explained as a combination of the metal calx and phlogiston. In the middle of the eighteenth century, chemists argued that the latter reaction was possible because of the affinity between phlogiston and metal calces. Some chemists went a step further to explain chemical affinities on the analogy of Newton's law of gravitation as a short-range attractive force between differently shaped atoms. However, because all assumptions about atoms and their shapes were purely hypothetical, most chemists ignored attempts to further explain chemical affinity on a deeper ontology of atoms and Newtonian forces.

THE CHEMICAL REVOLUTION

The reforms that Antoine-Laurent Lavoisier and his collaborators, Antoine François Fourcroy, Louis Bernard Guyton de Morveau, and Claude-Louis Berthollet, introduced into chemistry in the 1770s and 1780s have long been at the center stage of historians' interest. Historical interpretations of these events have varied considerably. Based not least on Lavoisier's own assessment that he had achieved a "revolution" of chemistry, earlier generations of historians long agreed that Lavoisier, with the support of his collaborators, introduced a deep structural transformation of chemistry that amounted to a relatively sudden and radical break with chemistry's past, that is, a scientific "revolution" (Guerlac 1961; Donovan 1988; Bensaude-Vincent 1993). More recently, several historians have renewed the arguments in favor of this view (Kim 2003; Crosland 2009; McEvoy 2010; Chang 2015). In contrast, other historians have voiced skepticism over the revolution model (Eddy et al. 2014). Some thirty years ago, Frederic L. Holmes pointed out that "a thriving investigative tradition existed in chemistry long before the revolutionary period." He further argued that Lavoisier and his group "cannot have overturned the science of chemistry as a whole, or have established a science for the first time," even though he retained the expression "chemical revolution" (Holmes 1989: 103, 107). Today, there is a broad consensus among historians of chemistry that "chemical revolution" does not mean that chemistry came into existence as a science only with Lavoisier's chemistry; rather, they agree that chemical science existed earlier in the eighteenth century as well as in pre-eighteenth-century alchemy.

Historians of chemistry also agree that Lavoisier and his collaborators introduced important novelties into eighteenth-century chemistry. Lavoisier introduced a new theory of combustion and calcination of metals by substituting the theory of oxygen and caloric for the traditional theory of phlogiston. He also introduced the modern concept of "simple substance" (or "chemical element" in this sense) and, with his collaborators, a new chemical nomenclature, based on an understanding of the chemical composition of substances. This group of chemists further reversed the classes of simple and compound substances. Substances previously considered to be relatively simple substances – air, water, acids, the metal calces, and alkalis – were now regarded as chemical compounds, while what were formerly regarded as chemical compounds such as metals, sulfur, and phosphorus were now defined as "simple substances" (chemical elements). To this impressive list of innovations others can be added, such as Lavoisier's oxygen theory of acidity; his theory of the states of aggregation; his redefinition of organic substances as compounds consisting of carbon, hydrogen, and oxygen; his contributions to the understanding of fermentation and respiration; and his balance-sheet method of supplementing studies of chemical reactions by routinely and quantitatively comparing the masses of the original substances with the masses of the reaction products (Donovan 1988; Holmes 1989).

However, there is no consensus about the question of whether these innovations ought to be interpreted as a "revolution" of chemistry in the sense of a sudden break with its past and introduction of new paradigms along with deep structural reorganizations. The alternative interpretation embeds the work of Lavoisier and his group into a long-term historical and communal context, highlighting continuities with their predecessors. On this alternative view, Lavoisier reaped the rewards of a century during which chemists extended their empirical knowledge, refined their methods, and restructured their understanding in a more gradual way. Even though Lavoisier introduced many innovations and challenged chemists on theoretical, methodological, and linguistic fronts, there was no chemist at the time who would have had serious difficulties in understanding these novelties. On the contrary, the adherents of the phlogiston theory could easily translate Lavoisier's theories into their own, and preserved many of their fundamental concepts and classificatory schemes. This fact has been explained by pointing to the shared conceptual framework of Lavoisier (and his group) and the generations of eighteenth-century chemists living before him (Klein and Lefèvre 2007; Klein 2015b).

THE ENLIGHTENMENT CONTEXT

The Enlightenment was a period when many literate members of society sought to promote progressive principles such as liberty, equality, toleration, and sociability, as well as the practical usefulness of knowledge. During the eighteenth century there was a widespread belief that these principles were best achieved and maintained through founding or reforming schools, academies, and universities in ways that provided students, rich and poor, with a well-rounded education. Particular emphasis was placed on obtaining a firm knowledge of the liberal arts and the natural world. In this context chemistry's longstanding study of natural substances and utility to trade, industry, artisans, agriculture, and medicine made it the perfect subject to include in curricula that sought to prepare aspiring minds for their future attempts to improve the economic and moral health of society.

The integration of chemistry into educational institutions required new forms of national and local support and regulation. Monarchs established professorships in chemical subjects at universities. Alternatively, at the time many universities were run by town councils which relied on them to train the professional classes and who wanted to attract as many good students as possible. As Enlightenment principles took stronger hold, councils also wanted to appoint professors who could help improve the financial well-being of the town. The professorships created by councils throughout the century satisfied all these conditions. The chemists of the University of Edinburgh's Medical School, for instance, were called upon to advise on sanitation, potable water, and other important public health issues. They also acted as advisors to the mining, linen, and ceramics industries that provided vital economic security for the Scottish Lowlands (Clow and Clow 1952). A similar situation existed for the chemists employed to teach at the mining academies that existed in Scandinavia, Central Europe, and further east in St. Petersburg (Schleiff and Konečný 2013).

Chemistry's association with enlightened notions of progress and improvement was significantly aided throughout the century in government corridors by the fact it successfully delivered workable solutions to important problems arising from the supply or manipulation of materials used to make munitions, coinage, fuel, and other commodities which were vital to maintaining military power in Europe and abroad in the colonies (Roberts and van Driel 2017). Enlightened monarchs, governments, and administrators increasingly sought chemists to advise them on topics ranging from waste disposal to agricultural productivity. German principalities facilitated the process through the practice of cameralism, that is, the professional training of government administrators who, among other essential subjects, were educated to understand the chemical arts (Meinel 1988; Klein 2015a; Klein 2020).

The creation of progressive chemical solutions depended upon the frequent circulation of new and reliable facts. During the Enlightenment there was a strong drive to collect and collate chemical knowledge into organized and easily accessible systems (Eddy 2008). From the 1740s onward, chemists were inspired by the immense success of Carl Linnaeus' multi-volume *Systema naturae*, which divided up the minerals, plants, and animals of the entire globe into a neat, hierarchal nomenclature of classes, orders, genera, species, and varieties (Klein and Lefèvre 2007). To make chemistry easier to understand, professors and textbook authors worked very hard to arrange all of the substances of chemistry into clearly delineated systems that jointly made sense to students and served as reference points for experimentalists seeking to determine which areas of chemistry needed to be clarified or expanded (Beretta 1993). Central to all these systems were affinity tables, which offered a summary of the most common chemical substances and their relative reactions, and various kinds of diagrams, which helped track the substances in compounds subjected to analysis and synthesis (see above).

When it came to the circulation of useful or new chemical discoveries and their applications, many chemists benefitted from the Enlightenment's principle of sociability, that is to say, a commitment to politely interacting with others with a view to reaching a deeper understanding of the world and the human condition. One of the most voluminous examples of this commitment was the *Encyclopédie,* published in numerous volumes from the 1750s to the 1770s. As noted above, it contained a plethora of articles that were filled with a wealth of chemical knowledge.

Enlightened readers who wished to learn more about the theories and practices of chemistry could consult one of the many chemistry textbooks and handbooks regularly published in most countries. Those who wanted to know about the most recent experiments could read the chemistry articles published in the journals of learned societies such as the Royal Society of London, the Académie des Sciences in Paris, and the Königliche Preussische Akademie der Wissenschaften in Berlin, all of which published articles and advertised prizes relevant to the social, industrial, and medical applications of chemical substances. Near the end of the century journals devoted exclusively to chemistry appeared as well, particularly in Germany, France, and Britain, where chemically literate readers could consult specialized journals. We already encountered Crell's *Chemische Annalen* and *Annales de chimie* above, but there were others such as Johann Bartholomäus Trommsdorff's *Journal der Pharmacie für Aerzte und Apotheker* and William Nicholson's *A Journal of Natural Philosophy, Chemistry and the Arts* (Klein 2007). The editors of these journals worked hard to offer different kinds of articles, book reviews, and letters to create a community readers who collectively helped to advance chemistry and its connections to the other sciences (Knight 2006).

Within the new world of chemical publishing there was sometimes a fine line between self-interest and social improvement, and between self-fashioning and selflessness. Even the most improvement-minded experimentalist had to put bread on the table. Such was the case of personalities such as the German protestant minister Jacob Christian Schäffer, who combined his knowledge of chemistry and natural history to make different kinds of paper from an assortment of materials derived from plants or animals that lived in his own locality. Whether making paper from wasp nests or sawdust, he worked diligently to publish his results so that they might be used by others and further his own desire to be known as a scientist worthy of recognition by experts based in universities or who were members of learned societies (Szalay 2019). When Schäffer's efforts are considered alongside other forms of new knowledge generated by other chemists, it becomes clear that by the end of the century there was a wealth of specialized information available for those who wished to use it to improve society via the production of commodified or purified substances.

THE PUBLIC IMAGE OF CHEMISTRY

Many people during the Enlightenment believed that they were living in the Golden Age of chemistry. By the end of the century the discipline could boast the establishment of new professorships in prestigious universities, the discovery of new substances, the invention of cutting-edge warfare technologies, the purification of water and foodstuffs, the creation of novel pharmaceuticals, the proliferation of sophisticated instruments, and the development of advanced industrial processes. Likewise, when compared to their seventeenth-century predecessors, eighteenth-century chemists believed that they had a much better understanding of the theories and practices that underpinned the creation of chemical knowledge.

Chemistry's positive image in the public eye was related in part to its appeal to the ever-growing reading public of Europe. Ever since the sixteenth century chemical substances had appeared regularly in works of literature. By the eighteenth century poets, playwrights, and novelists tantalized the reading public with stories featuring occult forces, mysterious elixirs, and experimental metaphors (Thompson 2017). Authors used their characters' knowledge of chemistry to support plots, either as enigmatic alchemists with unexplained powers or as arch-empiricists who wished to reduce humans into helpless piles of matter subjected to natural forces they could not control. In this context the gothic novelist Ann Radcliffe introduced female characters who used chemistry for the good of society. In *The Romance of the Forest* (1791), for instance, the character Madame La Luc used her own chemical laboratory to make medicine for villagers (Chandler 2006; Köhler 2013: 111).

Throughout the century essayists, clerics, and, indeed, some experimentalists presented erudite interpretations of the natural world, using chemistry to illustrate the divine designs they saw in the organization of minerals, plants, animals, and the cosmos. Some chemists, such as Joseph Priestley, moved beyond orthodox religion and advocated a form of Christian materialism in which chemistry played a central role in the resurrection of the dead in the end of days. These works were joined near the end of the century by authors, influenced by Romanticism and German *Naturphilosophie*, who sought to move beyond the mechanical view of nature (Knight 2004; Knight 2013; Knight 2016). The culmination of these movements occurred in novels such as Goethe's *Die Wahlverwandtschaften* (*Elective Affinities*) (1809a) and Mary Shelly's *Frankenstein* (1818), both of which explored the limits of reason by using chemical metaphors and theories.

The increasing availability of chemically prepared, useful materials caught the attention of sociable literati who believed it was important to learn more about chemistry on account of the role that it could play in improving the human condition. These and other pragmatically minded concerns also inaugurated a popular demand for public lectures in universities, aristocratic courts, homes, and the expanding urban spaces of the metropole. Public lectures of chemistry, supported by demonstration experiments, rendered chemistry the epitome of experimental science that demonstrated the transformability and productivity of nature. The price and quality of such lectures varied, but over time the bangs, sparks, and scents of chemistry impressed many different audiences and transformed it into a science of spectacle, raising its status and acceptability with the general public (Golinski 1992).

A key to chemistry's success in these events was the rising number of female participants and attendees. In Prussia, Martin Heinrich Klaproth's public chemical lectures attracted many women (Klein 2015a). In France and Switzerland, women such as Emilie du Châtelet, Marie-Anne Paulze Lavoisier, Claudine (Poulet) Picardet, and Albertine Necker de Saussure organized and conducted chemical experiments (Rayner-Canham 1998). During the last half of the century women frequented lectures on chemistry given in North America and Europe offered by traveling speakers such as the blind experimentalist Henry Moyes. They witnessed or facilitated experiments with air pumps and voltaic piles in private gatherings hosted by families and friends associated with salons or learned societies. They patronized chemistry lectures given in newly founded scientific organizations like Anderson's Institution in Glasgow and the Royal Institution in London.

The presence of girls and women at chemistry lectures was memorialized by a number of artists. Perhaps the most striking depiction was Thomas Wright of Derby's painting *An Experiment on a Bird in the Air Pump* (1768), which is now housed in the National Portrait Gallery in London. One of the best-known works of eighteenth-century British art, it depicts a group of observers standing around an air pump in a candle-lit room. Wright most likely painted it to reflect the scientific gatherings that took place in the homes of the members of the Lunar Society of Birmingham, England (Uglow 2011: 208–11). Two of the nine observers are girls, and one is a young woman. By this time the air pump was a standard instrument used in chemistry lectures. Demonstrators used it to create a vacuum inside a glass bell jar and conducted experiments inside it. The presence of female attendees in Wright's portrait speaks to chemistry's public status as a subject worthy of feminine attention.

Later in the century the caricaturist James Gillray captured a similar situation in his print *Scientific Researches* (1802). It placed women and a girl eagerly taking notes in the front row of a Royal Institution chemistry lecture. One such woman who attended the lectures was Jane Marcet (née Haldimand), the salonnière whose family had emigrated from Geneva to London. She was inspired to conduct experiments at home and went on to publish *Conversations on Chemistry, Intended More Especially for the Female Sex* in 1805. The book presented chemical concepts in clear prose and described experiments that could be conducted in a kitchen (Knight 1986). Marcet's knowledge of chemistry combined with the presence of girls in Gillray's print and Wright's painting draws our attention to the fact that some progressive parents took care to hire tutors and buy books so that their daughters could learn about chemistry (Leigh and Rocke 2016). During the 1790s the Midland-based Scottish industrialist James Keir even went so far as to write a chemistry instruction manual for his daughter, Amelie. Entitled *Dialogues on Chemistry between a Father and His Daughter*, it featured diagrams and explained the elements of chemistry and its associated forms of experimentation (Moilliet 1964; Eddy forthcoming).

The public perception of chemical substances was often subtle; they were a little-noticed part of daily amenities afforded to those living in large European cities like London, Paris, Prague, Vienna, and Berlin, locations that played an important role in setting the fashion, health care, and entertainment trends elsewhere. Take for instance Franz Anton Mesmer's popularization of hypnotism, the existence of which he explained by appealing to animal magnetism and a biochemical understanding of the human body. He first advertised its therapeutic effects in Vienna, the capital of the Austro-Hungarian Empire and a metropole with educated inhabitants whose perception of chemistry was influenced by its presence in the neo-humoral rationales offered by the medical professors teaching in its university and by the successful remedies offered by chemists promoting bathing in mineral springs in Central European spa towns such as Karlsbad (Karlovy Vary) and in monastic medicine cabinets maintained at abbeys such as those at Melk, Kremsmünster, and Pannonhalma (Vogel 2010). Politics eventually forced Mesmer to leave, but he took his ideas to Paris, a place which also boasted educated and affluent inhabitants who were familiar with the notion that chemistry could be used to improve health.

As intimated above, urban audiences were exposed to the wonders of chemistry by traveling lecturers, many of whom used their fame to sell instruments and books to make a living. But perhaps the most sensational impact of chemistry occurred in the theater and the art world, where painters used chemical substances to create new and brilliant blues, effervescent pinks and tropical oranges for portraits, landscape paintings, and even the backdrops of stage productions. As evinced in the striking metallic luster of the blue hues that seemingly leap off the canvas in Thomas Gainsborough's famous *Blue Boy* (ca. 1770), the chemical formulae of many eighteenth-century paints still attract the attention of scholars today (Lowengard 2008; Vankin 2018). Outside the elite salons of artists and collectors, chemists collaborated with theater companies, royal benefactors, city councils, and private citizens to create breathtaking pyrotechnic displays. These events ranged from increasingly sophisticated fireworks designed to sparkle across the skyline of Moscow to elaborate spotlights and dancing flames in London theaters (Werrett 2010; Golinski 2017).

In the late eighteenth century, chemistry's public image was also aided by the central role that chemical substances played in the Industrial Revolution and its reliance on coke for fuel, alloys for steam engines, and solvents for breaking down substances, many of which owed their genesis to chemically trained inventors and entrepreneurs (Musson and Robinson 1969). Chemical industries played a central role in the production of textiles, steel, alloys, and fashionable domestic goods. Textile factories financed research and development on substances such as dyes, mordants, and bleaches that were used to fix or remove colors from fibers. Mining firms worked with chemically trained metallurgists and engineers to improve steel and develop new alloys. Producers of domestic goods employed chemists to develop new, cheaper manufacturing processes for household necessities such as pottery, glassware, ceramics, soaps, oils, and candles.

The public image of chemistry was also aided by the everyday private world of useful substances that existed in many households. Prior to the nineteenth century, many chemical laboratories were located in private houses – most notably in apothecary's houses – and household members sometimes participated in coordinating and observing experiments (Klein 2008; Leong 2018; Werrett 2019). Add to this the fact that families and religious communities such as monasteries and convents often used chemical knowledge to make everyday household items like ink, bitters, preserves, and alcohol. Likewise, the ingredients listed in cookbooks were sometimes the same as those used in chemical experiments. Families collected and then sold domestic waste and old linen to venders who then transformed it into other marketable substances (Strasser 2014). In sum, the public image of chemistry was underpinned by the fact that it was a science that could be witnessed in everyday situations by householders living cheek by jowl in domestic communities.

The importance of chemistry to the ever-expanding world of global commodities delivered a constant supply of new products into the eager hands of middle- and upper-class consumers. For example, chemistry was employed to refine indigo grown in Asia into beautiful blue dyes used to make fashionable clothing (Kumar 2012), to transform essential oils from the tropics into fragrant perfumes or preservatives that prevented wooden instruments from drying out

(Sheller 2003: 43–5), and to mimic the glossy white porcelain of China (Klein 2014a). It also helped European physicians, colonists, and bio-prospectors to understand indigenous drugs such as cinchona bark that had been successfully used for centuries by the Indians of the Americas (Schiebinger 2004; Crawford 2014). Here the commercial success of chemistry alerts us to the numerous side effects that simultaneously worked against its public image. Substances like indigo and exotic oils were often acquired through global networks that depended upon the kinds of exploitation that were part and parcel of European imperialism. Within Europe, avant-garde cosmetics sometimes contained dangerous substances like lead compounds that harmed the body over time. The softer, whiter linen produced by factories relied on substances like bleach that spilled into rivers. The thick steel plates used to make the steam engines that powered much of the Industrial Revolution were made in coal furnaces that billowed noxious smoke into cities and the countryside.

For many of the problems created by industry and commerce, chemistry also offered solutions, which contributed to its public standing. The young environmental sciences of geology, meteorology, hydrology, and agriculture were founded on a chemical understanding of substances and theories of heat that many believed could be used to combat the ill effects of the climate, pollution, and soil deprivation (Eddy 2007; Golinski 2010).Those suffering from health hazards of the city increasingly flocked to spa towns where local medical experts lectured on the positive chemical effects of mineral water, effervescent tonics, cold baths, and clean country air (Eddy 2010). Within urban environments politicians and administrators turned to chemically literate experts when seeking advice on the kinds of legislation needed to promote beneficial solutions designed to protect residents from environmental health hazards (Albritton-Jonsson 2013; Tomory 2017). The fact that chemistry was seen as a solution to the problems that it had helped to create further reveals how the public often chose to concentrate on its many successes, a situation that explains how its reputation remained resilient in the face of the serious impact it continued to have on areas in which the chemical industries operated.

CHAPTER ONE

Theory and Concepts: *Transformations of Chemical Ideas in the Eighteenth Century*

URSULA KLEIN

During the eighteenth century, chemists questioned and revised fundamental beliefs about the ultimate structure of matter. At the same time, they also introduced novel chemical concepts and theories based on accumulated experience acquired in chemical workshops and laboratories. In this chapter, we will follow the decline of speculative philosophies of matter in eighteenth-century chemistry as well as the introduction of new chemical concepts and theories, which eventually led to the "Chemical Revolution" in the last third of the eighteenth century.

THEORIES OF THE ULTIMATE STRUCTURE OF MATTER: ELEMENTS, PRINCIPLES, AND ATOMS

In the early eighteenth century, the concepts of elements, principles, atoms, and corpuscles were part of "philosophies" (later, "theories") about the ultimate structure of matter, which had a long tradition going back to antiquity. It was widely recognized that the world was furnished with a great number of different kinds of material substances. In the ancient philosophical tradition, this observation prompted the question of how the different kinds of naturally occurring substances were created in nature and what caused their various properties. The Greek atomists postulated that nature made the different kinds of natural substances by juxtaposing invisibly small particles, or atoms, of a universal matter that differed merely in size, shape, and motion (or rest), but not in any other quality. Hence all common natural substances were defined as mechanically assembled compounds whose properties were ultimately caused by the three fundamental mechanical properties (size, shape, motion) of the simple atoms. By contrast, the Aristotelian philosophers argued that the natural substances were "generated" from the four "elements" earth, air, water, and fire, which carried the four "qualities" dry, wet, cold, and hot; "generation" was regarded as a genuinely creative process that brought the different natural substances into existence. In the process of generation, the qualities of the elements merged with one another with the result that natural substances were fully homogeneous entities, the same in all their parts. Compared to the simple elements, never found in pure form in nature, natural substances were also called "mixts," because they originated from different elements. As mixts were fully homogeneous, however, their constituting elements no longer existed in their original physical state; they were rather "potentially" contained in the mixt. The distinctiveness of natural mixts was determined by the different proportions of the four original elements as well as by an immaterial "form," a concept that has no equivalent in modern chemical thinking. It was the immaterial form that rendered a part of matter a particular natural kind or species.

Paracelsianism, the predominant alchemical tradition in the early modern period, also defined the naturally occurring substances as homogeneous mixts that were generated from simple elements and principles, whose number ranged between three and five. In this framework of thinking, the more abstract Aristotelian concept of generation and matter-form dualism took on naturalized, gendered meanings. For Paracelsians, "generation" was reinterpreted as a process that was analogous to the generation of living organisms. It was initiated by the insemination of the passive elemental matrixes or wombs (earth and water) by the active principles or seeds (mercury, sulfur, and salt). In continuation of the Aristotelian tradition, the elements and principles were conceived as carriers of sets of qualities – dry, wet, colored, combustible, and so on – that caused the properties of the various kinds of homogeneous natural mixts. Influenced by Renaissance philosophies, Paracelsians also differentiated between the visible corporeal shell and the internal, invisible spirits of the homogeneous natural mixts, in analogy to body and soul of living beings (Klein 1994a; Klein 1994b).

The philosophy of elements and principles as well as early modern atomism and the mechanical–corpuscular theories referred to the bedrock of matter, and they shared the same reductionist ontology: the belief that behind the mundane natural world there was another, deeper level of natural things that were simple and stable and that caused the great variety of naturally occurring substances belonging to the mineral, vegetable, and animal kingdom. Throughout the eighteenth century these speculative natural philosophies remained controversial. Even though the majority of eighteenth-century chemists included them in their teaching and chemical textbooks, they were agnostic, or even skeptical, towards them, and in their research they often focused on more specific theoretical questions. In so doing they followed a broader movement that promoted explanatory pluralism and thus sought to identify non-reductionist, intermediate causes of phenomena (Gaukroger 2010; Chalmers 2012a; Chalmers 2012b). By the end of the century, Antoine-Laurent Lavoisier gave voice to this trend when he characterized all kinds of questions about the simplest elements or indivisible atoms as "metaphysical" ones that must be excluded from the science of chemistry (Lavoisier 1965: preface, 24).

STAHL'S THEORY OF ATOMS AND CHEMICAL PRINCIPLES

In the early eighteenth century, the medical professor and chemist Georg Ernst Stahl framed an influential theoretical agenda for rethinking the relationship between philosophies about the ultimate structure of matter and chemical experience. Hélène Metzger, and more recently Ku-Ming Chang, have shown that Stahl adopted the mechanical–corpuscular theory for explaining chemical change. In many of his publications, he presented chemical change as the result of the movement, collision, and juxtaposition of corpuscles with different shapes (Metzger 1930; Chang 2002; Newman 2014). However, Stahl was anything but an unambiguous follower of Boyle or Newton. His mechanical–corpuscular explanations of chemical change remained on a general level and abstained from detailing how differences of the shape of corpuscles related to observable effects. What is more, like many of the early-eighteenth-century French academic chemists (Principe 2001; Kim 2003; Boantza 2013a), Stahl also integrated the philosophy of simple chemical principles into his theoretical thinking, and in his theory of the origin and constitution of natural substances he eschewed strictly mechanical–corpuscular explanations.

Building on the work of the chemist Johann Joachim Becher, Stahl postulated that there were four simple chemical principles, namely water plus three "earthy principles" (in analogy to the three Paracelsian principles salt, mercury, and sulfur), which were the simplest components of natural substances. Each of the four simple chemical principles consisted of characteristic indivisible atoms, which possessed a certain size and shape as well as a set of characteristic non-mechanical qualities such as color and combustibility. It was, in particular, the non-mechanical character of the simple chemical principles that Stahl considered to be essential for getting to grips with chemical diversity. The shapes of atoms and corpuscles, he stated in this context, were an "occult quality," and "we will never acquire sufficient knowledge" about them. Hence, the mechanical–corpuscular philosophy, he further pointed out, left unanswered the question of how a particular property of a natural substance was caused by a distinct shape of its corpuscles (Stahl 1720: 50–1). By contrast, he believed that the philosophy of chemical principles was more illuminating in this respect.

Like the Paracelsian concept of principles, Stahl's concept of chemical principles differs profoundly from the modern concept of chemical element. The modern concept refers to a large number of simple, undecomposable substances existing in nature in a free state. Hence, chemical elements in the modern sense are common chemical substances that can be stored in vessels, used in the laboratory, and manipulated in experiments like any chemical compound. By contrast, in the spirit of the chemical–philosophical tradition, Stahl's simple chemical principles were defined as the hidden causes of common natural substances, and the number of these principles was small. There was thus a fundamental ontological difference between simple chemical principles and the naturally occurring compound substances. However, like the leading French chemist of his time, Wilhelm Homberg, Stahl also believed that hidden causes can be revealed by observation and experiments. Therefore, he struggled to find reliable empirical methods to identify the simple chemical principles. His basic assumption was that the hidden "causes" must resemble their "effects" and therefore became known through their effects. Knowledge about a simple chemical principle, Stahl thus stated, must be "a posteriori [knowledge] based on its important and particular effects" (Stahl 1720: 53).

Following Becher, Stahl regarded knowledge embedded in the mineralogical tradition as one way to shed light on the simple chemical principles, and he thus considered the traditional classification of minerals into four classes – earths and stones, metals, combustible minerals, and salts – as a guideline to clarify their qualities. This method was based on the assumptions that the four chemical principles were distributed unequally in the mineral kingdom – not all minerals were made up from all four principles, and they also contained different proportions of the principles – and that the four mineralogical classes were natural classes. All stones and earths – such as granite, sandstone, gems, gravel, clay, lime, or chalk – were relatively heavy, dry, solid, hard, dense, incombustible, and vitrifiable when heated. Hence, Stahl postulated that these minerals contained a large proportion of a "primitive" or "vitrifiable earth" – corresponding to the Paracelsian principle of salt – that caused these characteristic properties (Stahl 1720: 119–22). In metals, the "mercurial earth," corresponding to the Paracelsian principle of mercury, was predominant, which rendered the metals ductile and malleable. Sulfur, bitumen, coal, and other combustible mineral substances contained a large proportion of the "inflammable principle" or "sulfuric earth," corresponding to the Paracelsian sulfur, which caused their combustibility, color, and so on. Stahl also gave the name "phlogiston" to this inflammable principle (see below). The fourth mineralogical class, which included common salt, vitriol, saltpeter, and so on, contained substances made up from a relatively large proportion of the principle of water, which rendered these minerals soluble.

The second empirical way to examine the principles was experimental decomposition. Based on his corpuscular theory and knowledge about reversible chemical operations (see below), Stahl believed that natural substances were heterogeneous compounds that actually contained the chemical principles in the form of stable building blocks. A complete analysis of natural substances, which he also called "mixts," thus meant their decomposition into the preexisting simplest components. However, Stahl warned his readers that chemical experimentation had not yet accomplished the resolution of mixts into their most simple components. We know "a posteriori," he stated, that we "nowhere come across the principles of mixts in a pure and unmixed state, isolated from their true state of combination among each other." If a simple principle was separated from a substance, he pointed out, it "immediately associates itself anew in another way" (1720: 31–2; 1730: 4, 12).[1] Hence, Stahl emphasized time and again that when chemists spoke of the simple chemical principles such as sulfur, mercury, and salt, obtained through chemical decomposition, they were not referring to the naturally occurring substances named sulfur, mercury, and salt. Instead, he pointed out, there was only a certain "analogy or similarity" between the former and the latter (Stahl 1718: 70). Likewise, he observed that chemists' principles of earth and water were not the common substances named earth(s) and water, but rather a single pure water and a single pure earth hidden in the common natural substances.

Stahl's skepticism concerning chemists' experimental achievements and knowledge about simple principles highlights an issue that would become a vexing epistemological problem for his followers. If the simple chemical principles could not be obtained experimentally in a pure, isolated state, then there was no way to provide empirical evidence for their existence and for their distinct nature and number. As a consequence, eighteenth-century chemists added numerous variations to Stahl's original theory. Even Stahl's inflammable principle ("phlogiston"), which was widely accepted in the chemical community, was subjected to skeptical questions and modifications (see below).

STAHL'S HIERARCHICAL MATTER THEORY

As we have seen above, in the Aristotelian and Paracelsian tradition, all naturally occurring substances were conceived as mixts generated from a small number of elements or principles. Even though Stahl adopted much of this view, including the term mixt, he eschewed the Aristotelian and Paracelsian assumption of the homogeneity of mixts, as well as matter-form dualism. The ancients, Stahl asserted, had believed that mixts were the outcome of the "total mutual penetration of the principles," which was due to the fact that they were ignorant about possibilities of "artificial resolutions and combinations" of substances (1720: 12). The scholastic "conceptus materialis and formalis" (matter-form dualism), he further pointed out, had no empirical meaning. Nor was the naturalized reinterpretation of "form" intelligible, such as Johann Kunckel's "spermatis" regarded as a "generative or mixing force" that determines "the particularity of each genus [of substance]" (1718: 57). Presenting empirical cases of decomposition and subsequent recomposition of chemical substances, Stahl gave the concept of mixt a new meaning: a "mixt" was no longer a homogeneous body but a compound that contained the simple principles in the form of bodily components or atoms.

This was a significant step away from the traditional concept of mixts and a significant move toward the modern concept of a chemical compound. A closer look reveals, however, that Stahl's concept of mixt was not as modern as it may seem, because Stahl kept the traditional philosophical meaning of the components of mixts. Stahl defined the components of mixts as ultimate chemical principles that were small in number and belonged to a more fundamental ontological level than the perceptible natural mixts. As the ultimate chemical principles neither existed in a free state in nature nor could be easily separated from the mixts, Stahl had no direct empirical knowledge about them. This implied that there was also no direct empirical access to the generation and composition of mixts. By contrast, the modern concept of chemical compound refers to substances that can be both produced from and decomposed into the ordinary chemical substances that are its components.

However, Stahl also provided a theoretical platform that allowed later generations of chemists to bypass some of the ontological and epistemological problems connected with the concepts of simple principles and mixts. He made a distinction between mixts and more compounded substances, and between the ultimate principles and more proximate, compounded principles. Building on Becher's theory, he posited that there were three major classes of naturally occurring substances. First were the simple "mixts" that consisted immediately of the simple chemical principles; these kinds of natural substances were extremely difficult to decompose, and there was no agreement among chemists about their composition. Second, there were the more compounded "compounds" that consisted of two or more mixts; the substances belonging to this class could be decomposed by means of chemical operations, and their components, the mixts, could be isolated experimentally in the form of chemical laboratory substances. The third class were the "decompounds" containing compounds as their proximate principles; the substances belonging to this class could be easily decomposed as well. Becher had also introduced the more complex level of "superdecompounds," but Stahl rarely used the latter concept (Stahl 1720: 6). All of these concepts could also have an atomistic meaning in certain contexts.

Stahl emphasized that the number of chemical species that belonged to the class of mixts was small and that all of these substances belonged to the mineral kingdom. It was not yet clear, however, which kinds of mineral substances were mixts and which were compounds or decompounds. The chemically purified metals, Stahl observed, may be mixts, as they could be decomposed into a particular metal calx and phlogiston, which could not be further decomposed. Likewise, chemists had transformed sulfur into sulfuric acid and phlogiston, which they interpreted as a decomposition of sulfur (Klein 1994a; Kim 2003). However, chemists had neither accomplished the isolation of pure phlogiston nor could they be certain that metal calces and sulfuric acid were simple principles. On the contrary, Stahl pointed out, it was more likely that the latter were still mixts. The class of "compounds" was considerably larger than that of the mixts. Stahl stated that most "minerals extracted from the earth are compounds consisting of mixts" or even "decompounds" (Stahl 1720: 9). For example, the metal sulfides were compounds or decompounds consisting of sulfur and a metal, for chemists had decomposed them into sulfur and a metal.

Stahl's hierarchical matter theory was particularly well received in the French chemical community. The belief that there was an increasing complexity of composition ranging from mixts to decompounds, or even superdecompounds, implied the assumption that there was a corresponding chain of principles ranging from the four simple principles to the more compounded or proximate principles. Unlike the simple principles, the more compounded principles could be isolated from naturally occurring substances and subjected to further experimental studies. They were just common chemical substances. What they shared with the simple principles was their characterization as property-conferring components in view of the higher level of composition. By the mid-eighteenth century, Stahl's theory of a hierarchy of composition provided a theoretical framework for chemists to focus chemical analysis on the more proximate principles of chemical compounds (Holmes 1989; Klein and Lefèvre 2007). The term "principle" in the sense of proximate principles began to merge with the term "chemical component." The next section illuminates the meaning of the modern concept of chemical component along with new concepts such as "chemical compound" and "chemical composition," which have roots in experimental and practical contexts.

NEW CONCEPTS AND THEORIES: CHEMICAL COMBINATION, CHEMICAL COMPOUNDS, AND AFFINITY

In the course of the seventeenth century, chemists' empirical knowledge about the possibilities of chemical operations had significantly accumulated and changed in character. Whereas "dry operations" by means of fire such as distillations and smelting processes had long predominated in chemical workshops and laboratories, "wet operations" such as dissolutions of substances in mineral acids and subsequent precipitations were carried out more frequently (see Chapter 2 in this volume). Seventeenth-century apothecaries, in particular, produced a wide spectrum of chemical remedies by means of acidic dissolutions and subsequent precipitations, which yielded new kinds of salts. These salts could again be decomposed into their original ingredients. Chemical practitioners and academic chemists who were engaged in teaching and in learned inquiry repeated, varied, and further extended such kinds of "reversible chemical operations" and published them in their textbooks and academic memoirs (Klein 1994a; Klein 1994b). Based on this practice, chemists of the Paris Academy of Sciences developed a new chemical theory centered on the concepts of chemical combination, compound, and affinity. This theory intimately coupled the understanding of chemical change with the theory of chemical composition.

In the early eighteenth century most chemists believed that natural substances, including those belonging to the mineral realm, naturally grow and ripen. Hence, they assumed that in the bowels of the Earth the base (ignoble) metals were slowly transmuted into silver and gold. In accordance with this view, they regarded numerous cases of chemical change they carried out in the laboratory as artfully accelerated imitations of the substances' natural change. Fire, they believed, enhanced the properties of material substances, and thus accelerated their natural ripening. Transmutation was a special case of the enhancement of the properties of a single metal in which its properties were thoroughly changed, to the extent that a new kind of metal emerged. Extraction of the subtle essence of a natural substance was another type of chemical change, regarded as a process that eliminated the substance's gross corporeal parts and thus brought its nobler qualities to the fore (Holmes 1989; Klein and Lefèvre 2007). Both the concept of transmutation and that of extraction of essences considered chemical change as change of the properties of a single substance.

In the decades around 1700, Nicolas Lémery and Wilhelm Homberg introduced a new chemical theory of salts based on empirical knowledge about the preparation of salts. The theory interpreted the composition of chemically prepared salts in terms of tangible components into which these salts could be decomposed and recomposed, and it further added a corpuscular explanation of this view (Holmes 1989; Klein 1994a; Kim 2003). As Frederic L. Holmes observed, this new understanding of salts "was far from a simple or direct empirical generalization. To arrive at it, chemists had to accept a proposition that was difficult for them to comprehend; that is, that a substance could still be present within a combination even though its characteristic properties had disappeared" (Holmes 1989: 37). In 1718, Etienne François Geoffroy took the next step, publishing a table of "chemical rapports" that grouped together almost all known chemical operations that could be interpreted as reversible chemical operations (Geoffroy 1718).

Geoffroy's table embodied a group of new chemical concepts that referred to chemical substances used in the laboratory and their chemical reactions. It avoided any hypothesis that involved atoms, corpuscles, and the deeper causes of chemical change (Klein 1994a; Klein 1995; Chalmers 2012a). The table presents two types of chemical change and, by implication, a related concept of chemical compound and composition. First, there were chemical combinations of two kinds of substances in the sense of a symmetrical interaction, or reaction, of these two substances and their grouping together in an elective way according to chemical "laws." Here the term "laws" referred to the constant, intrinsic relationships ("rapports") between pairs of substances. Such laws varied in degree, depending on the kinds of substances undergoing chemical change, and determined the tendency of these respective substances to combine with each other. The result of chemical combination was a "chemical compound" in which the two combining substances persisted in the state of the compound's "chemical components." These chemical components could again be obtained in the state of free substances through the decomposition or "chemical analysis" of the compound.

The second type of chemical reactions presented in Geoffroy's table was displacement reactions: When a substance had a stronger "rapport" with one of the two components of a chemical compound, it displaced the second component and took its place. It thereby decomposed the chemical compound and simultaneously created a new one. In the middle of the eighteenth century the majority of chemists were convinced that chemical decomposition, or chemical analysis, predominately took place in the form of displacement reactions, governed by laws of chemical affinity. In contrast, the predominant understanding of analysis in the alchemical period had highlighted fire as the instrument of decomposition, which meant that a single agent acted upon a passive substance causing its decomposition. Moreover, as the majority of alchemists regarded natural substances as "mixts" in the sense of homogeneous materials, "decomposition" implied the re-creation of the original physical state of the constituents of the homogeneous mixt, rather than mere decomposition into pre-existing physical components.

The system of chemical concepts presented in Geoffroy's table is basic to modern chemistry, but their historical development has received little attention until recently. Many historians and philosophers assumed that it was implicitly part of ancient atomism and the early modern mechanical corpuscular theories (Duhem 1902; Newman 2006). Clearly, the concepts of chemical combination, compound, composition, analysis, and so on implied a certain mechanical understanding of chemical reactions and composition: chemical substances were regarded as relatively stable building blocks that could combine, were preserved in the chemical compound in the state of its actual physical components, and could be recovered and rearranged in subsequent reactions. There are, however, important distinctions between the chemical concepts embodied in Geoffroy's table and the concept of a compound and composition in the tradition of atomism. The "chemical compound" can be both made up of and broken down into the chemical substances that are its components. Thus, the entities that move, combine, and separate are always tangible chemical substances available in the laboratory. Moreover, the "chemical compound" is defined as the result of chemical combination caused and directed by law-like "rapports" (later "affinities") between substances.

By contrast, in atomism and the seventeenth-century mechanical–corpuscular theories the entities making up a compound are invisibly small corpuscles, or even indivisible simple atoms, and their ability to combine is explained exclusively in terms of mechanical qualities such as motion, size, and shape of corpuscles. Based on the idea of different corpuscular structures, which Robert Boyle called "textures" of compound corpuscles and defined as the cause of the different kinds of natural substances, the early modern mechanical corpuscular theories could be adapted to the alchemical concept of transmutation in the sense of change of the properties of a single substance and its transformation into a new kind of substance. Far from highlighting symmetrical reactions between different substances and displacement reactions as the most important type of chemical change, Boyle and his followers used the concept of texture of compound corpuscles to explain transmutations of single substances: in transmutation the texture of a compound corpuscle disintegrated into particles of the universal matter ("prima naturalia"), which then reconfigured to form a different texture of a new kind of compound corpuscle making up a new kind of chemical substance (Klein 1994a; Newman 2014).

Geoffroy's table has long been celebrated for the introduction of the theory of chemical affinity (Kim 2003). Some historians have argued that "rapports" actually meant Newtonian attractive forces and that Geoffroy was bringing into French chemistry the corpuscular theory presented in the Queries of Newton's *Optics* (Crosland 1963; Thackray 1970). By contrast, Alistair Duncan has interpreted the table as a systematization of empirical knowledge without any underlying theory (Duncan 1996). Overlooked in these interpretations is the fact that, in his commentary to the table, Geoffroy claimed to have formulated a "theory." But in his theory he avoided any hypotheses concerning the ultimate structure of matter. Geoffroy's theory worked on the level of chemical substances and the relationships (rapports) between them, not on the level of invisibly small corpuscles and attractive forces between corpuscles. It was independent of any theory of the ultimate structure of matter, including the theory of elements and principles. Even so, it included an entire system of new concepts, and was both explanatory and predictive. As Alan Chalmers has pointed out, it can be interpreted as knowledge of intermediate causes just as Newton's theory of gravitation in the *Principia* can. Neither gravitation on the one hand nor chemical combination, composition, and affinities on the other necessarily required further explanation on the deeper level of ultimate causes (Chalmers 2012b). As we will see in the following sections, in the second half of the eighteenth century Stahl's hierarchical matter theory provided a bridge between the system of chemical concepts embodied in affinity tables and the theory of chemical elements.

STEPS TOWARD THE MODERN CONCEPT OF CHEMICAL ELEMENTS

In the course of the eighteenth century, chemists' reasoning about chemical principles and elements changed profoundly. This change was not least promoted by the growing importance of the analytical method and the affinity tables, along with the system of new chemical concepts they embodied. When in the middle of the century the chemist Pierre Joseph Macquer summarized chemists' understanding of simple principles, the theoretical shift became visible. Having described the proper method of performing a chemical analysis and its supplementation by the re-synthesis of the decomposed substance, Macquer proposed the following definition of a simple chemical principle: "But this analysis and this decomposition of bodies is limited. We can push it only up to a certain point, beyond which all of our efforts are useless. Whatever the manner we reach this point, we always come to a halt through substances that we find to be unalterable and that we cannot further decompose, and which function like barriers that we cannot trespass" (Macquer 1753: 2). This was a novel definition of "chemical element," although Macquer kept the term principle as well as the idea that there were just four simple principles. Macquer moved the concept of simple principle (or element) further away from its traditional philosophical context, and instead linked it with the system of chemical concepts embodied in Geoffroy's affinity table. Taking Stahl's hierarchical matter theory as a theoretical bridge, he assumed that chemical analysis proceeded in several steps. It first yielded the most compounded proximate principles of a chemical compound that could be further subjected to chemical analysis, which yielded less compounded proximate principles; only in the last step of a whole series of chemical analyses did chemists encounter a limit of decomposition. The chemical substances obtained in this last analytical step, Macquer argued, had to be regarded as the simplest chemical principles. Thus, the analytical method became the arbiter for the identification of simple principles.

Macquer presented his new definition of a simple chemical principle in the context of his explanation of the analytical method, but he did not draw general theoretical conclusions from his definition. Instead, this devoted follower of Stahl posited that there were just four simple principles, namely air, water, fire or phlogiston, and earth (Siegfried and Dobbs 1968). The assumption that "air" was a principle of natural substances was new. It relied on recent experiments with different kinds of air (later, gases) and, in particular, on Stephen Hales' (1727) *Vegetable Staticks* (see Chapter 2 in this volume). Macquer further proposed that in its free state the principle air was identical with common air. The element air, he stated, was "the fluid that we are breathing continuously" (Macquer 1753: 4). According to Macquer, simple air existed both as a naturally occurring "element" and as a simple "principle" or component of naturally occurring substances. Likewise, water existed both in the state of a combined principle and as a free "body so well known that it is almost useless to give a general idea of it" (Macquer 1753: 6). Based on Stahl's idea that released phlogiston was the body of fire (see below), Macquer further stated that the principle of phlogiston corresponded with the element "fire." The element "fire," he further pointed out, was the matter of the sun, heat, fire, and flames. Only in the case of the principle of earth, the corresponding free element was still unknown, as it was "very difficult and even impossible to get the principle of earth entirely free from all other substances" (Macquer 1753: 9–10). Hence, Macquer no longer made a deep ontological distinction between simple principles and naturally occurring substances.

LAVOISIER'S CONCEPT OF ELEMENTS AND PRINCIPLES

Macquer had proposed a modern, analytical definition of simple principles, but he had kept the idea that there were just four such simple chemical substances. The man who relinquished the latter aspect of the ancient philosophy of elements and principles was Antoine-Laurent Lavoisier. "If we apply the term *elements*, or *principles of bodies*, to express our idea of the last point which analysis is capable of reaching," Lavoisier posited, "we must admit, as elements, all the substances into which we are capable, by any means, to reduce bodies by decomposition" (Lavoisier 1965: preface, 24). Thus, chemical elements were not only defined as the substances obtained in the last step of chemical analysis and that cannot be further decomposed, as Macquer had claimed. In addition, Lavoisier also stated that there were many such indecomposable, or "simple substances," and that these simple substances often existed in a free state. In their Méthode de nomenclature chimique, Lavoisier and his collaborators listed fifty-five simple substances. Chemical elements were now understood to be just simple chemical substances – and nothing else. They were a particular class among the multiplicity of chemical substances. Lavoisier was well aware that this was an important intellectual move away from the chemistry of Stahl and his followers. The theories about the "four elements" and about the "simple and indivisible molecules," he observed, had stimulated purely metaphysical discussions (Lavoisier 1965: preface, 24). For Lavoisier, such discussions had to be excluded from the new science of chemistry. This was a clear break with the past, but one that had been prepared by generations of chemists before Lavoisier. It was neither a sudden rupture, a "revolution," nor was it an absolute conceptual novelty that would have erased all ideas associated with the ancient concept of principles. On the contrary, Lavoisier, who highlighted the role played by language in science, kept the ancient term "principle" for elements that entered chemical compounds as components rather than existing in a free state. With that move he also kept an important idea of "principlism" (Chang 2012), that is, the idea that chemical elements or principles were property-conferring components of chemical compounds.

Hence, Lavoisier substituted the name "oxygen" for "pure air," derived from the Greek words "oxys" (sharp) and "genes" (born of), because it was the component or principle of chemical compounds that conferred the property of acidity to them. He defined oxygen as "an element common to them all [acids], which constitutes their acidity," in short as an "acidifying principle." He further introduced the term "azote," again drawn from Greek language for "not-living," for the principle contained in what we know as nitrogen gas, the second major part of the atmospheric air; the name highlighted "its known quality of killing such animals as are forced to breathe it." Likewise, he pointed out that "hydrogen" was "the generative principle of water," and that the matter of heat or "caloric," another simple substance, conferred the property elasticity to the gases (Lavoisier 1965: 22, 51, 52, 65, 89). Following Robert Siegfried, Hasok Chang has interpreted the rise of the modern concept of chemical composition ("compositionism") as an event that was closely linked with the decline of the theory of principles and eventual abolishment of "principlism" in the "chemical revolution" (Siegfried and Dobbs 1968; Siegfried 2002; Chang 2012; Chang 2015). He thus presented "compositionism" and "principlism" as theoretical alternatives that mutually exclude each other. By contrast, Lavoisier's new concept of element (or "simple substance") can also be interpreted as the result of a theoretical reconstruction and partial integration of "principlism" into "compositionalism."

THE THEORY OF PHLOGISTON

As intimated above, Lavoisier's ideas operated in a heterogeneous intellectual environment. It took time for chemists to evaluate his claims and to examine how they related to other important chemical theories. This was especially the case for the theory of phlogiston. To fully understand phlogiston's rise and subsequent decline within the theoretical landscape of the eighteenth century, we must revisit the ideas of Georg Ernst Stahl. Beginning in 1697, with the publication of his *Zymotechnia Fundamentalis*, Stahl elaborated a unified theory of combustion and calcination of metals, today known as the "theory of phlogiston," which was widely accepted in European chemistry. Stahl's concept of phlogiston was embedded in his theory of principles, which postulated that all common natural substances were composed of three or four simple principles, which caused their physical and chemical properties. "Phlogiston" was just another name for "sulfuric earth," or the "principle of combustion," or the "inflammable principle." Stahl argued that phlogiston caused inflammability, combustibility, the tendency of metals to corrode and become a calx, as well as the color, smell, and taste of natural substances (Stahl 1718: 77–83).

According to Stahl, in the combustion of substances their phlogiston was released and immediately transferred into the surrounding air, where it acquired the "movement of fire," observable as flames. Hence, Stahl posited that phlogiston was the "proper matter of fire" or "bodily fire," and what was commonly called "fire" was nothing else than free phlogiston in violent motion, which presupposed air as a receiver and supporting medium. As phlogiston in the free state of fire expanded quickly and dispersed in the air, it could not be perceived by the senses (Stahl 1718: 78–9). Heat was extremely dispersed phlogiston that was also unobservable. In order to begin the release of phlogiston, a body had to be ignited with fire, that is, phlogiston in motion. The quickly moving particles of fire pushed the fixed particles of phlogiston, set them in motion and thus helped them to rip apart their ties. Clearly, this explanation of ignition and combustion borrowed from the corpuscular philosophy.

A novelty of Stahl's theory of phlogiston lay in his claim that the corrosion of metals could be understood as basically the same kind of chemical process as combustion. For example, when iron rusted it released phlogiston into the air; it was thereby transformed into iron calx, consisting of the primary earth and the mercurial earth contained in the original iron. Iron could be recovered when it was heated with charcoal, which contained a large proportion of phlogiston. For this extension of the phlogiston theory, Stahl assembled numerous observations from mines, foundries, and the metallurgical crafts, where metals and metal alloys were frequently submitted to reversible chemical operations (Klein 1994b). His examples of decompositions and recompositions of metals, in which phlogiston was first released and then added again, demonstrates the importance of reversible chemical operations for his theory and for eighteenth-century chemical concepts in general.

The particular ontological status and nature that Stahl ascribed to phlogiston – its definition as a causal principle and the assumption that it was the matter of fire dispersing in the air – had important epistemological consequences. It had never been possible, Stahl conceded, to study phlogiston "in and for itself, independent of being mixed and compounded with other materials, and hence to detail its own properties" (1718: 79). At the same time, however, Stahl also aimed to strengthen the resemblance of phlogiston with ordinary chemical substances available in the laboratory. He pointed out that phlogiston could be transferred from one substance to another, and that it behaved in chemical reactions like ordinary chemical substances. He also emphasized that it was a "bodily principle" (106, 131) and that proofs for its existence must take into consideration "both measure and weight." Phlogiston was extremely light, however: "in each pound [of ordinary sulfur] there was hardly one ounce of phlogiston" (329). It did not escape Stahl's attention that the metal calces were heavier than the original metals, but he did not endeavor to explain this observation.

Another novelty of Stahl's phlogiston theory was its extension to all kinds of naturally occurring substances. It was not only minerals that contained phlogiston, as Becher had believed; vegetable and animal substances contained this principle as well. Stahl also believed that there was a full cycle of phlogiston in nature. Metals could be recovered from their calces by means of the vegetable substance charcoal, which contained a large proportion of phlogiston. As Stahl had demonstrated in an early experiment, it was also possible to recover common sulfur from sulfuric acid by means of a vegetable oil. Hence, Stahl claimed that chemists' experiments demonstrated that phlogiston could be transferred "from the vegetable and animal to the mineral things" (83–4). The transfer of phlogiston also went in the other direction, from minerals to plants and from plants to animals, so that a complete circle resulted. During their growth, plants assimilated phlogiston from the soil and, more importantly, from the air, where it had been released through the combustion and corrosion of substances. Finally, the animals' fats stemmed from the vegetable oils and fats they consumed.

Both the intellectual origins and the further development of Stahl's phlogiston theory are complex matters. On the one hand, the meaning of phlogiston depended on the ancient concept of hidden causal principles in the tradition of Aristotelian and Paracelsian philosophies, which Stahl amalgamated with the mechanical corpuscular philosophy. Hence, phlogiston was a hypothetical entity, and Stahl conceded that it could not be separated from the mixts in the state of an isolated chemical substance that could be stored in vessels. On the other hand, Stahl's phlogiston theory was also embedded in knowledge about reversible chemical operations acquired in the context of pharmaceutical and metallurgical practice and chemical experimentation. Hence, Stahl also endorsed the new analytical method that demanded to complete the chemical analysis of a substance by its resynthesis. Additionally, in view of the subsequent development of the phlogiston theory, another feature of Stahl's theory ought to be emphasized: Stahl created a conceptual link between phlogiston and fire and heat; phlogiston in the free state of motion was the matter of fire and heat. However, Stahl did not further explore the relationship between phlogiston, fire, and heat, as Macquer would do some decades later.

PHLOGISTON AND THE MATTER OF HEAT

By the mid-eighteenth century, chemists generally agreed that phlogiston behaved in chemical reactions just like any ordinary natural substance (Kim 2003; Boantza 2013a; Klein 2015b). It was involved in chemical reactions that were caused by chemical affinities between pairs of different kinds of substances. This made the problem of separation and isolation of pure phlogiston all the more urgent. Like Stahl, Macquer defined phlogiston as a simple principle, but he also highlighted its problematic status. Chemists had only "confused and inexact ideas of this inflammable principle," he observed (Macquer 1766: vol. 2, 202), which was not least due to the fact that it had never been isolated in a free state, according to standards that chemists had recently applied even to the most subtle kinds of air (later gases). When the second edition of Macquer's *Dictionnaire de chymie* was published in 1778, the problem of isolating phlogiston was still unsolved. The inflammable "vapors" studied in the new field of pneumatic chemistry now became candidates for isolated phlogiston. The "inflammable gas" (later "hydrogen"), in particular, which was released when zinc or iron was dissolved in sulfuric acid, might be almost pure phlogiston, Macquer observed. The "electrical matter" was another candidate for pure phlogiston, which remained hypothetical as well (Macquer 1778: vol. 2, 196). Macquer did not find an empirical but rather a theoretical solution to the problem, when he tackled the question of how phlogiston related to heat and light.

In the middle of the eighteenth century, the relation between phlogiston and matter of heat was one of the major theoretical problems. Macquer also took up this second challenge. Based on thermometry and contemporary physical studies of heat, he distinguished between the fixed matter of heat, that is, phlogiston, and the free matter of heat. His systematic comparison of the two led to a vexing problem. The free matter of heat dispersed in all directions, penetrated all bodies, and caused expansion, especially in the volume of fluids. Volume expansion through the matter of heat was used in thermometry to measure temperature. Furthermore, Macquer argued that all substances, including air, would be in the solid state without the existence of the matter of heat, because solidity and fluidity, including the "fluidity" of air, depended only on the quantity of free matter of heat penetrating a substance. All of this meant that the free matter of heat was in violent motion, elastic, and could not be combined with a substance and thus transformed into the fixed state. Yet, phlogiston was exactly defined as the fixed matter of heat. Moreover, the matter of heat had apparently lost its elasticity in the fixed state of phlogiston, because metals and many combustible substances were perfectly solid substances. As early as 1753, Macquer pointed out that this was a puzzling problem, and that there was "no satisfying answer" to it (Macquer 1753: 15). Clearly, phlogiston had become a problematic substance long before Lavoisier began to attack it.

In 1778, just a year after Lavoisier had published his discovery that combustion, calcination, and reduction of metals involved oxygen and launched his first attack on the phlogiston theory of combustion (Guerlac 1981), Macquer had found a solution to his problem: he dismissed the concept of matter of heat and replaced it by a mechanical theory. Heat was now defined as "the motion of vibration or oscillation" of the parts of bodies, and all bodies and substances were susceptible to this kind of motion (Macquer 1778: vol. 2, 197). Concerning phlogiston, Macquer pointed out that studies of light had shown that light could be manipulated in ways that resembled chemical analysis. In addition to dispersion and reflection, it could also be decomposed and recomposed. If light could be decomposed like ordinary substances, he hypothesized, it might also be possible that it could be combined with material substances. As a consequence, he proposed that phlogiston was nothing else than the fixed matter of light. He endorsed this proposition with additional empirical arguments, the most important one being that plants needed light for growing and that they lost color when light was absent. Hence, their phlogiston stemmed from light; it was the fixed matter of light.

At the same time, Macquer revised the phlogiston theory of combustion and calcination. Like many other chemists of his time, he accepted Lavoisier's discovery that oxygen was involved in combustion, without simultaneously dismissing phlogiston (Boantza 2013a). According to Macquer's revised phlogiston theory, in combustion and in calcination of metals the vital part of air (Lavoisier's "oxygen gas") combined with the combustible substance taking the place of phlogiston. Hence, phlogiston was set in motion and released into the air. In its free state, phlogiston was the matter of light in motion, that is, it engendered light, visible as flames. A metal calx had a greater weight than the metal, because it had combined with vital air, and the weight of combined vital air corresponded with the weight of air consumed from the atmospheric air during combustion; by contrast, phlogiston was either imponderable or it had a very low weight. The reduction of a metal was the inverse operation of calcination. Under favorable conditions, phlogiston displaced fixed vital air from its combination with the metal calx and took its place.

In Macquer's final phlogiston theory, the understanding of heat had radically changed, and the relationship between phlogiston and light had been clarified. The revised phlogiston theory of combustion took oxygen and weight relationships into account, and it was integrated into chemical reasoning with affinities showing that combustion, calcination, and reduction could be understood as displacement reactions. Phlogiston remained a substance involved in combustion, calcination, and reduction as well as an important property-conferring component of chemical compounds. It still was the principle of combustibility, color, smell, taste, ductility of metals and so on, and there was also the possibility to link it with studies of electricity. Macquer's revised phlogiston theory unified many different chemical areas.

THE "CHEMICAL REVOLUTION": LAVOISIER'S THEORY OF OXIDATION AND CALORIC

Antoine-Laurent Lavoisier has long been regarded as having initiated a radical rupture in chemistry – a "chemical revolution" – and thus created the foundations of modern chemistry (Guerlac 1961; Donovan 1988; Bensaude-Vincent 1993; Brock 1993; Siegfried 2002; Crosland 2009; McEvoy 2010; Chang 2015). In the last two or three decades, however, many historians of chemistry have challenged this view, uncovering patterns of continuity and gradual change taking place over many decades (Holmes 1989; Klein and Lefèvre 2007; Boantza 2013a; Eddy et al. 2014; Klein 2015b; Lefèvre 2018). Historians of chemistry have long regarded Lavoisier's oxidation theory of combustion and calcination as an important, or even the crucial part, of his theoretical innovations. As Maurice Crosland has observed, the "Chemical Revolution of the late eighteenth century consisted essentially of combustion being explained by the addition of oxygen rather than the removal of phlogiston. This has been seen as the 'paradigmatic shift' of a scientific revolution in the familiar Kuhnian sense" (Crosland 2009: 93). However, Lavoisier's final theory of combustion made two combined assumptions: that oxygen is added to the combustible substance, and that the imponderable matter of heat is simultaneously released from the "oxygen gas." It thus rested on two unequal pillars: on quantitative experiments involving oxygen and on the more traditional theory of the matter of heat. The former showed that a distinct part of the atmospheric air, identical with Joseph Priestley's dephlogisticated air, was absorbed in combustion and calcination and increased the weight of combustible substances and metals, respectively. The latter explained typical phenomena accompanying combustion: the production of heat and flames. The next section provides an overview of the second, often neglected pillar of Lavoisier's theory of combustion, which shows that there are significant similarities between Lavoisier's theory and Macquer's revised phlogiston theory of 1778.

LAVOISIER'S THEORY OF CALORIC AND CHANGE OF STATES OF AGGREGATION

Like Macquer (before 1778), Lavoisier believed that the matter of heat, which he renamed "caloric," was an imponderable, invisible substance, which moved quickly, penetrated all vessels, and possessed elasticity, which he defined as force of repulsion. Unlike Macquer, however, he assumed that caloric could be fixed in substances without losing its elasticity. Caloric, he argued, combined chemically with most substances due to its chemical affinity with them, and it conferred elasticity to the compounds of which it was a component. Gases were elastic because they contained a large proportion of caloric. Furthermore, building on the physical concept of "heat capacity" and on the observation that during changes of the state of aggregation a substance's temperature did not alter, Lavoisier posited that changes of the state of aggregation were not physical processes, as we conceive of them today, but rather chemical changes. When a solid substance became liquid, it combined chemically with caloric, and when liquids changed into the state of gases, they combined with yet another proportion of caloric, which saturated them with caloric. Changes of the state of aggregation could be reversed when temperature decreased and caloric was released from the compound. Lavoisier thus defined all liquids and all gases as chemical compounds containing caloric as a component.

Hence, strictly speaking, Lavoisier pointed out, a new name for a substance would be necessary when it changed its state of aggregation. However, as this was inconvenient in view of linguistic conventions, he restricted himself to renaming just the gaseous substances. In order to highlight their compounded nature, he introduced double names for them (Lavoisier 1965: 50). For example, what is today designated simply as oxygen he named "oxygen gas," which signified that the gas was a chemical compound consisting of oxygen and caloric. Accordingly, Lavoisier's "oxygen" was a substance that had not yet been isolated chemically in a free state. Hence, Lavoisier's theory postulated a plethora of simple substances that were as hypothetical as the rejected phlogiston. But his theory of gases and states of aggregation shared yet another feature with the phlogiston theory: Lavoisier kept the idea of property-conferring principles (see above). Caloric functioned as a principle conferring repulsion or elasticity to the gases. Hence, Lavoisier stated that the "elasticity [of gases] depends upon that of caloric, which seems to be the most eminently elastic body in nature. Nothing is more readily conceived, than that one body should become elastic by entering in combination with another body possessed of that quality" (Lavoisier 1965: 22).

A vexing problem of this theory was the relationship between the repulsive force of caloric and its affinity, or force of chemical attraction, with other chemical substances. How did the chemical affinity between caloric and chemical substances overcome the effects of the repulsive force of caloric, which caused volume expansion of a heated substance? In corpuscular terms, volume expansion meant that the particles of a substance and the particles of caloric moved away from each other rather than combining each other. When heating continued, more caloric was added to the substance, which had the effect that the substance suddenly changed its state of aggregation. It was in this single "moment" of the change of the state of aggregation, Lavoisier argued, that the force of chemical attraction was stronger than the force of repulsion, and the particles of caloric combined with the particles of the substance. This was a plausible mechanism, but it was an ad-hoc explanation that could not have been applied to other cases of chemical change. Lavoisier conceded that the science of chemistry was still "in a state of imperfection," and that its most difficult part was the theory of chemical affinities (Lavoisier 1965: preface, 20–1).

LAVOISIER'S THEORY OF COMBUSTION AND CALCINATION

In Lavoisier's theory of combustion and calcination, caloric fulfilled the same explanatory functions as phlogiston did in the phlogiston theory. Based on the theory that gases were chemical combinations of a substance, such as oxygen, with caloric, Lavoisier explained the production of heat during calcination and combustion with the release of caloric (or matter of heat) – in analogy with the traditional explanation with the release of phlogiston. In combustion and calcination of metals the oxygen contained in the oxygen gas combined with the combustible substance or the metal, respectively, and simultaneously caloric was released from oxygen gas causing an increase of heat and, in many cases of combustion, flames. While Macquer had found a solution concerning the question of how the heat and the light of flames related to each other, this question remained open in Lavoisier's theory.

Like Macquer's revised phlogiston theory of 1778, Lavoisier described combustion, calcination, and reduction as displacement reactions. For example, in the calcination of metals the affinity between the metal and oxygen was greater than the affinity between the oxygen and caloric contained in the oxygen gas. Hence the metal displaced caloric from its combination with oxygen, and took its place. Metal oxides were heavier than metals because oxygen was added, and their increase of weight corresponded with the weight of the consumed oxygen gas; caloric was imponderable. Oxidation could also be reversed by means of substances that had a great affinity with oxygen. For example, in the reduction of metal oxides, carbon displaced oxygen from the metal oxide and combined with it as well as with free caloric forming the gas carbon dioxide.

However, oxygen had another explanatory function in Lavoisier's theory, which is often overlooked and was also analogous to the explanatory function of phlogiston: oxygen was a property-conferring principle. According to Lavoisier, the oxides resulting from combustion were acids. For example, the reaction product of the combustion of sulfur was sulfuric acid, that of phosphorus was phosphoric acid, that of carbon (charcoal) was carbonic acid, and so on. Hence, Lavoisier argued that oxygen was a principle that conferred acidity to chemical compounds. In the case of metal oxides, which were not acids, Lavoisier stated that they were not fully saturated with oxygen but rather "intermediate substances, which though approaching to the nature of salts, have not acquired all the saline properties" (Lavoisier 1965: 79). In this case, he regarded oxygen as a color-conferring property, as the color of the metal oxides was often proportional to the degree of their oxidation.

What Lavoisier dismissed in his theory of combustion and calcination was not the general concept of principles but rather a very particular principle: phlogiston. What is more, he replaced this particular principle with another one: caloric, which was both imponderable, like phlogiston, and a principle that confers the property of elasticity. Historians of chemistry have long argued that Lavoisier's theory of oxidation was a novel theory, and that its novelty lay in his substitution of oxygen for phlogiston. However, they often ignored the crucial role that caloric played in Lavoisier's theories. Furthermore, as the discussion of Macquer's revised phlogiston theory has shown, the defenders of the phlogiston theory could easily integrate the discovery of oxygen into their own theory. In Great Britain, famous chemists such as Joseph Priestley and Richard Kirwan defended the phlogiston theory (Boantza 2013a). In the eyes of the late phlogistonists, the controversial point was not oxygen, but rather the question of how properties such as combustibility, color, smell, and ductility could be explained if phlogiston was dismissed. Instead of dismissing traditional reasoning with principles, Lavoisier introduced oxygen and caloric as alternative property-conferring principles.

The majority of late-eighteenth-century chemists believed in the existence of invisible and imponderable substances such as phlogiston, caloric, and the matter of light. In 1778 Macquer replaced the concept of matter of heat by a mechanical theory of heat, but he still believed in the concept of a matter of light, which he implemented into his revised phlogiston theory. Lavoisier dismissed the concept of phlogiston, but he still believed in the imponderable matter of heat or caloric. Both the late Macquer and Lavoisier explained combustion, calcination, and the reduction of metals as displacement reactions. They thus succeeded in ordering these reactions into the paradigmatic class of reactions highlighted by affinity tables. But while Macquer explained combustion as the displacement of phlogiston, contained in the combustible substance, by oxygen,

Lavoisier explained it as the displacement of caloric, contained in the oxygen gas, by the combustible substance.

Lavoisier introduced an important theoretical novelty: combustion and calcination go hand in hand with the addition of oxygen. However, the presence of caloric and property-conferring principles in Lavoisier's theories should not be seen as a minor oddity that was soon corrected by further work. Rather, it should be seen as indicative of the extent to which Lavoisier absorbed and built on the tradition of property-conferring principles and the phlogiston theory, and transformed it into a theory of the same kind. In this sense, these later versions of the phlogiston theory and Lavoisier's theory of combustion and calcination were mirror-image theories.

CHAPTER TWO

Practice and Experiment: Operations, Skills, and Experience in Eighteenth-Century Chemistry

VICTOR D. BOANTZA

INTRODUCTION

Eighteenth-century chemists employed instruments, operations, and experimental setups similar to those used for centuries by alchemists, medical chemists, and metallurgists. Distillation devices, balances, and furnaces remained central to their practice (Holmes 1988: 18–19). The implication that experimental practice tends to change more gradually than theoretical developments is commonplace in the history of all sciences. Aptly labeled "the fairest flower" in the "history of the experimental sciences," chemistry was no exception (Bensaude-Vincent and Stengers 1996: 38). The chemical laboratory "evolved very gradually, at a pace measured not in years but in generations of chemists. Operations were refined, the designs of furnaces improved. New reagents were added, and chemical indicators supplemented the basic tests of color, odor, and taste." This continuity was disrupted in the latter third of the century, when Antoine Laurent Lavoisier and his collaborators re-established chemistry on new theoretical, linguistic, and instrumental grounds. Often referred to as a chemical revolution, this period of reformation began in the early 1770s and peaked around the publication of Lavoisier's *Traité élémentaire de chimie* (*Elements of Chemistry*) in 1789. However, there were also important continuities: from the turn of the eighteenth century to the chemical revolution, relying on traditional instruments and techniques, chemists built an "increasingly dynamic investigative enterprise," oriented around the study of acids, bases, metals, and salts (Holmes 1988: 19).

Nicolas Lémery's *Cours de chymie* was one of the most popular chemistry textbooks of the late seventeenth and early eighteenth centuries. It went through multiple editions and was translated into several languages (Partington 1962). In France, where its last edition appeared in 1757, it remained authoritative until the publication in 1749 of Pierre-Joseph Macquer's *Elémens de chymie théorique*. Lémery's work belonged to a well-established French textbook tradition that emphasized practical and pharmaceutical goals, but he also dedicated considerable space to mineral chemistry and to descriptions of operations involving acids, alkalis, metals, and salts (Kim 2003: 55). His *Cours de chymie* paid limited attention to theory, especially when compared to the space devoted to chemical applications, recipes, substances, reactions, and equipment. In the twelfth edition, for instance, which spanned close to 800 pages, the "principles of chemistry" comprised a brief chapter at the outset, followed by a much lengthier discussion of the "proper furnaces and vessels for chemical operations." The rest of the book was divided into three parts, dealing with mineral, vegetable, and animal matters. At around 400 pages, the section on minerals was longer than the other two combined (Lémery 1724).

Discussing distillation, Lémery explained that "Chymists, in making the [fire] *Analysis* of mixt Bodies, have met with five sorts of Substances, they therefore concluded that there were five *Principles* of natural Things, *Water, Spirit, Oil, Salt,* and *Earth*." In this elemental pentad, water and earth were considered passive; the other three principles were seen as active. Lémery here echoed popular views that identified the products of fractional distillation with combinations of the four Aristotelian elements – earth, water, air, and fire – and the three Paracelsian principles – salt, sulfur, and mercury – named after the Swiss medical alchemist Theophrastus von Hohenheim (Paracelsus). Spirit, Lémery specified, designated mercury, whereas oil was "called *Sulphur* by reason of its inflammability … [and was] said to cause the Diversity of Colours and Smells." Salt was a "fixt, incombustible Substance, that gives Bodies their Consistence … [and] causes the Diversity of Tastes." In line with contemporary perspectives, Lémery highlighted the role of analysis, defining chemistry as "an Art that teaches how to separate the different Substances which are found in mixt Bodies" or "those things that naturally grow and increase, such as Minerals, Vegetables, and Animals." Lémery admitted that "the *five Principles* are easily found in *Animals* and *Vegetables*, but not so easily in *Minerals*," which included "the Seven Metals, Minerals, Stones, and Earths" (Lémery 1720: 1–3, 5).[1]

Lémery's definitions and distinctions invoke some of the main organizing concepts used in early-eighteenth-century chemistry and their relation to experimental practice. We can identify the importance accorded to distillation and to the elementary constituents of matter. The emphasis on direct experience and qualitative empirical data is likewise significant, based on the chemist's familiarity with the distinctive appearance and sound of reactions alongside the colors, smells, textures, and tastes of substances (Roberts 1995). Early-eighteenth-century chemists relied on classificatory schemes like the four elements, the three principles, and the three kingdoms of nature. Even the distinction between animate and inanimate matter is tacitly present in his remarks on chemical method. Although generally popular and selectively used by prominent chemists of the period (including Robert Boyle and Herman Boerhaave as well as Lémery), explanations based on particles and/or Newtonian forces were largely irrelevant to eighteenth-century chemical practitioners. Lavoisier ultimately dismissed all speculations about "the number and nature of elements ... [as] metaphysical," cautioning that "if, by the term *elements*, we mean to express those simple and indivisible atoms of which matter is composed, it is extremely probable we know nothing at all about them." In 1789 Lavoisier redefined chemical elements by replacing the quest for ultimate and metaphysical entities with the pragmatic and provisional reality known to chemists: "any substance ... must be considered as simple in the present state of our knowledge, and as far as chemical analysis has hitherto been able to show" (A.-L. Lavoisier 1790: xxiv, 177).

MATERIAL AND OPERATIONAL REPERTOIRE

Marcellin Berthelot once mused that "chemistry creates its own object. This creative faculty, akin to that of art, distinguishes it essentially from other natural or historical sciences, the object of which is given in advance and is independent of the scientist's will and action" (1876: 275). These words of a late-nineteenth-century author could be read as underscoring the productive role of chemistry with a nod to industry. They could also be read as the reflections of a seasoned experimentalist, keenly aware of chemistry's empirical intricacies and their complex relations to theoretical precepts, an awareness that induced this chemist to liken his profession to an "art." Both readings would be correct, but would underscore different nuances. I choose the latter for the image it evokes: of the chemist immersed in a continuous balancing act, explicitly and implicitly engaging the realms of nature and artifice. Chemists both study and create their objects of inquiry, often simultaneously. Knowledge of chemical objects and the circumstances of their production are often inextricably bound. Chemistry has always been a conceptual–practical hybrid, embodying in a general but essential sense, as Berthelot's words suggest, both artisanal and scientific dimensions.

This is not to say that chemical experiments and practices were always directly informed by conceptual guidelines or carried out strictly to further theoretical knowledge. The meaning of premodern "experiments," which commonly blended aspects of science, craft, technology, and industry, is broader than its modern counterpart. Fruitful efforts to unpack such premodern relations and forms of knowledge include accounts of chemistry as an "impure science," histories of "materials and expertise," "compound histories," and examinations of the "sites of chemistry" and the shifting notions of early modern chemical "laboratories" in relation to experimental practice, trade, and commerce (Klein 2008; Bensaude-Vincent and Simon 2009; Klein and Spary 2010; Perkins 2013; Roberts and Werrett 2017).

To understand the "actions" and "creations" of eighteenth-century chemists, their intimate involvement with their objects of production and study, as alluded to in Berthelot's striking image, we begin by surveying their repertoire of operations and the instruments and techniques they used to carry them out. With the rise to prominence of the new physical sciences in the aftermath of the scientific revolution of the seventeenth century, influential eighteenth-century chemists sought to carve out an independent epistemic and technical space for their occupation. They distinguished between the abstract, quantitative, and mechanical concerns of the physicists and their own empirical, qualitative, and unique chemical ways of inquiry. While physicists might explain effervescence, for instance, as the result of increased motion of submicroscopic particles, chemists would attribute it to particular tendencies of tangible substances to react and combine. They would subsequently trace the subtle qualitative transformations at hand to identify the relationships between diverse substances under various conditions.[2] Such disciplinary tensions were at the core of Gabriel François Venel's influential article on chemistry that was published in 1753 in Diderot and d'Alembert's *Encyclopédie*. Venel explained that the "principles of [mixed bodies] … are connected by a link different from the one governing the formation of aggregates or relationships of mass. The first can be broken down by mechanical as well as by chemical means. The second can be separated by chemical means alone." This was why, Venel concluded, "chemists are also good physicists" (1753: 413, 415).

Versions of such theoretical distinctions carried over into the experimental realm. Chemists divided most procedures into those involving solid materials ("dry" operations) and those based on fluids and solutions ("wet" operations). Mechanical operations were mostly carried out in the preparative stages of substances and were usually followed up by more chemically reactive processes. *Triturations*, for example, denoted the physical breakdown of solids through the use of grinders and mills. Two other dry mechanical operations were *attrition*, or the rubbing of one body against another, and *corning*, the coarse grinding of a solid substance. Wet mechanical procedures included *decantation*, the separation of a liquid suspended above a solid residue (usually the result of precipitation) by pouring the liquid off while carefully retaining the solids, and *elutriation*, the separation and purification of a mixture of granular solids with water by decanting, straining, and washing. *Filtration* meant, as it does today, a mechanical–physical separation of a liquid from a particulate solid by passing the liquid through a porous material like cloth or paper. When filtration was accomplished by passing the liquid through coarser media, like a hair sieve, it was called *collature*. Two operations, mainly performed on organic matter, were *expression*, the isolation of a fluid component from solid or semisolid matter by squeezing the material in a press, and *maceration*, a semimechanical process through which a solid sample could be softened by soaking it in a liquid. *Decrepitation* implied a process marked by the crackling sound produced when certain crystals were heated and thought to decompose by losing their water content.

Chemical operations of the dry kind normally entailed the application of heat, as in the heating to a high temperature of mineral solids, including metals. When such heated samples acquired a whitish color and a brittle consistency, the process was identified as *calcination*. During *deflagration* substances were made to burn vehemently; when the process produced a substantial release of heat, light, and noise, it was identified as *fulmination* or *detonation*. Whereas the roasting of ores to rid them of impurities was termed *torrefaction*, the restoration of a metal to its metallic state was known as *revivification*. Wet chemical processes comprised a more extensive category. *Coagulation* stood for the conversion of fluids into a solid form; *precipitation* meant the formation of a solid within a solution that would sink to the bottom of the vessel. *Dephlegmation* implied the removal of water from a solution, usually of acid or alcohol. *Ebullition* designated the agitating, bubbling action of a liquid. *Extraction* indicated the separation of a substance from others by using solvents. The separation of soluble from insoluble solid substances, known as *lixiviation*, was achieved by soaking or boiling a mixture of solids, then removing the resulting solution that contained any soluble material from the solid. *Infusion* referred to the extraction of chemical substances by soaking them in a solvent, usually hot water. *Digestion* had a wide range of meanings, from a broadly defined indicator of chemical activity or "inner motion," in contradistinction to "external" mechanical motion, to a slow process of organic decomposition, to the continuous application of heat to a substance without boiling it, usually attained in open vessels (Eklund 1975: 20–45).

Anatomists had traditionally relied on sharp scalpels to dissect the bodies and tissues of humans, animals, and plants. Fire was the chemist's scalpel: the main instrument used to decompose, transform, and purify matter. George Starkey, one of the best-studied early modern chemists or alchemists, whose work influenced Boyle, Newton, and others well into the eighteenth century, styled himself a "Philosopher by Fire." This label was nonetheless closely associated with experimental practice, as "it encapsulates the centrality [Starkey] gives throughout his laboratory work to testing his thoughts and theories by accessing the phenomena of the natural world as exhibited in chymical trials" (Newman and Principe 2002: 117). Lémery called chemistry "*Pyrotechnia* ... signifying the *Art* of *Fire*; for in effect it is by *Fire* we bring all Chymical Operations to pass" (Lémery 1720: 1). Because in the Aristotelian tradition fire was deemed the most active of the four elements, heat was considered a key agent in the generation, destruction, and perfection of matter. In their quest for the ultimate constituents of bodies, chemists turned to the traditional method of separation – the application of heat through burning and especially distillation – that became known as "analysis by fire" (Debus 1967; Holmes 1971: 131).

A significant subset of wet chemical reactions was related to *distillation*, the process of separating the components or substances from a liquid mixture by taking advantage of the different boiling points of the components to condense their vapors sequentially and separately. The five principles mentioned by Lémery had been modeled after the five fractions commonly obtained from most organic matter. When the collecting vessel was located above the heated vessel, the distillation was known as *per ascensum*; when it was the other way around, it was *per decensum*. *Rectification* meant the refining of a substance by one or usually multiple distillations. *Cohobation* was a common way of performing repeated distillations or any cyclic process whereby liquids were vaporized and recondensed. Chemists distinguished between *elixation*, the lengthy boiling or stewing of a substance, and *coction* or *decoction*, which called for a lengthy application of a more delicate form of heat. *Purification* denoted the general rendering of a substance free of other substances, often as a result of distillation. Similar to its modern usage, *solution* designated a liquid in which one component, the solute, was dispersed in another, the solvent (Forbes 1970; Eklund 1975: 20–45).

Eighteenth-century chemists employed three basic types of distillation apparatus (Figure 2.1). For general-purpose distillations they used the *alembic*, which consisted of three parts: the body or cucurbit, the neck, and a condensation vessel called the head. Most three-part alembics were made of glass. The length of the neck connecting the cucurbit to the head was related to the subtlety of the vapors reaching the head: longer necks were able to condense subtler vapors, whereas a shorter neck allowed the collection of heavier, less-volatile fractions. For lengthier reactions the standard apparatus was the *pelican*, a reflux device that kept the reactants in a constant vaporization–condensation cycle, by returning part of the condensed vapors from the head back to the cucurbit via two curved beaks. The operator had to carefully monitor the reaction until the amount of liquid collected in the cucurbit was deemed rich enough in the intended product. The newly formed product would then be

FIGURE 2.1 Various distillation retorts, pelicans, alembics, furnaces, and other chemical apparatus. From Nicolas Lémery, *Cours de chymie, contenant la manière de faire les opérations qui sont en usage dans la médecine ... Nouvelle édition, revue, corrigée & augmentée* (Paris: d'Houry, 1757), plates 1, 2, 5, and 6. Courtesy of HathiTrust.

separated by simple distillation from the leftover reactants. Lavoisier employed pelicans in his laboratory. For distilling heavier, less-volatile liquids, the typical choice was a *retort*, a vessel consisting of a spherical flask topped with a long neck pointing downward. The liquid to be distilled was placed in the vessel and heated, while the neck acted as a condenser along which the vapors could flow to a collection receptacle. The principal movement of vapor in a retort was therefore lateral rather than vertical. A major difficulty in using retorts, besides the inevitable risk of breakage, was the challenge of introducing substances into them at the beginning of the operation. Distillation, especially of the kind involving multiple fractions, was a sensitive process that required a great deal of skill and experience. The heat had to be carefully controlled and adjusted to the nature of the substances, their physical and chemical properties, as well as to the size and make-up of the apparatus (Eklund 1975: 9–11).

PLANT CHEMISTRY: TRADITIONS AND PRINCIPLES OF ANALYSIS AND CLASSIFICATION

Throughout the long eighteenth century the French Royal Academy of Sciences was at the forefront of European scientific activity. From its inception in 1666, chemistry had taken pride of place in the Academy. By the 1770s and 1780s it had become the institutional focal point of the chemical revolution on the European Continent, and Lavoisier and several of his close collaborators were among its members. This prominent tradition affords a valuable perspective on the development of chemistry, especially in the first and last thirds of the century. Physician Samuel Duclos and apothecary Claude Bourdelin were the first two chemists appointed in 1666. With Bourdelin's assistance, Duclos established and ran the Academy's chemical laboratory and set the research agenda for decades to come. Duclos had initially proposed the experimental study of the principles of natural "mixts," such principles broadly understood as the ultimate constituents of matter, for which he envisioned the use of both fire analysis (distillation), and solution analysis (solvent extraction). Bourdelin rejected the latter method: solvents had been controversially associated with metaphysical alchemical quests for universal all-powerful dissolving agents like the alkahest. Instead, Bourdelin performed and meticulously recorded vast numbers of distillations, mostly of plant matter, many of which were conducted as part of the Academy's "natural history of plants" project. Disputes over the meaning and validity of chemical analysis, combined with political and financial conflicts, resulted in the ultimate failure of the project.

Duclos shared earlier concerns, most notably expressed in Boyle's *Sceptical Chymist* (1661), that fire could not decompose matter into its simplest parts. Duclos also worried that any method based on differential degrees of heat was inaccurate because some constituents of a mixture might have boiling points that were so close together that distillation could not separate them. Yet instead of abandoning the practice, he prescribed its judicious use in combination with solution analysis. Duclos reckoned that a genuinely skilled chemist should be able to draw on his first-hand experience to track and interpret subtle qualitative transformations during analysis. No axiomatic rules or generalizable laws like those of the physicists, he thought, could replace the chemist's intimate experiential knowledge. The botanist Denis Dodart joined the Academy in 1673 and shortly thereafter took over the history of plants project. Dodart acknowledged the severe limitations of fire analysis, but shared Bourdelin's mistrust of solvents.

The lack of a demonstrable correspondence between plant matter and its distillates plagued analytical distillation practices. While apparently dissimilar plants could be generally reduced to the same five principles, the distillation of specimens that were similar – anatomically, medicinally, in terms of provenance, or otherwise – yielded fractions that were empirically different. Without a stable definition of chemical elements or composition, the chemists failed to reach consensus. Duclos hoped to use plant analysis, by fire as well as by solvents, to shed general light on the nature of matter. Bourdelin and Dodart prioritized the medical qualities of plants, for which distillation, although limited, was still regarded as useful. Personal rivalries ensued, and Bourdelin's single-minded efforts to compare the qualities of distillation products resulted in a forbiddingly detailed catalog that eluded convincing interpretation (Stroup 1990; Boantza 2013a).

At the turn of the century, the Academy welcomed a new cadre of chemists. The experimental natural philosopher Wilhelm Homberg was elected member in 1691. Three years later, apothecary Simon Boulduc was appointed. Nicolas Lémery, whom we have already met, and Etienne François Geoffroy were elected in 1699. Geoffroy's younger brother, Claude Joseph, joined their ranks in 1711. Homberg and Boulduc set out to reassess the experimental findings of their predecessors. For two decades, Homberg was the Academy's most active and influential chemist. Informed by trends in experimental physics, he conducted chemical experiments with air pumps and made extensive use of burning lenses to create concentrated heat to induce chemical reactions. A few decades earlier, Duclos had employed burning mirrors in experiments on the calcination of metals. Later in the century, Lavoisier would use a large lens to obtain carbon dioxide by burning a diamond. Joseph Priestley discovered oxygen, which he called dephlogisticated air, by deploying a smaller lens to heat a sample of "mercurius calcinatus" (mercuric oxide). E.F. Geoffroy, Macquer, and others also used burning instruments in chemical experiments. These inquiries constituted a key driving force in the conceptualization of fire as a chemical substance, especially in the first half of the century (Kim 2003; Lehman 2013; Boantza 2017).

The contrast between Duclos and Bourdelin was echoed to some extent in the approaches of Homberg and Boulduc to chemical analysis and the relations between chemistry, experimental physics, and medicine. The two factions can be seen as emphasizing different facets of chemistry, its function, and its epistemic role. Homberg and Duclos considered chemistry a means for inquiring into the general attributes of matter. The more traditional Boulduc and Bourdelin saw it primarily as a useful tool for procuring medicinal substances. Ironically, it was Boulduc's conservative focus on the medical virtues of plants that spurred a renewed interest in solution analysis that eventually undermined distillation methods. The issue turned on the way the chemists viewed plants as their objects of study. Reviewing fire analysis, Homberg targeted oils as one of the most distinctive constituents of plant matter. He examined the "fetid oils that were found at the end of each analysis," but "the great differences in the taste, odor, and consistency of [these] oils suggested to him that they might be further resolvable into simpler constituents," which he eventually identified as "water, salt, and earth." Despite concluding that Bourdelin's analyses failed to experimentally establish a discernable chemical difference between the substances obtained from an edible and a poisonous plant, like Duclos, Homberg still regarded fire analysis as a valuable source of proximate knowledge. Like Dodart before him, Homberg maintained that distillation, supplemented in special cases by other procedures, could produce viable if non-definitive knowledge about plant composition (Holmes 1971; 1989: 65–8, 73; Stroup 1979).

Boulduc applied Bourdelin's distillation procedures to the study of the root of the ipecacuanha plant, long known for its emetic powers, but soon became disenchanted with the method. Instead of fire analysis, he searched "for analogous and convenient solvents in order to extract each of these parts and to perform afterward experiments in order to ascertain in which of them its virtue, whether emetic or purgative, mainly resides" (Holmes 1989: 68–9). Focused on isolating the medically active ingredients, he used distilled water and alcohol as solvents. These extractive operations were hardly novel, but Boulduc broke new ground by prescribing their use in a more generalized fashion. He subsequently tried salt of tartar (potassium carbonate) solution and distilled vinegar. By emphasizing the isolation of medically *active* ingredients instead of ultimately *simple* constituents, Boulduc changed the criteria for a successful extraction, which in this case resided in the efficacy of the products as emetics or purgatives. The principles of mixts, according to Boulduc, should be recognizable by their medical effects. While the violent nature of fire and strong heat tended to destroy those virtues, the products of gentle solution retained medicinal (and other) qualities characteristic of the original material. Thus, solution methods satisfied Boulduc's twofold concern with the utility of chemical research and the qualitative permanence of principles in the extracts. The traditionalist had thereby popularized new methods: as part of his long-term plan to analyze all purgatives, Boulduc advocated a move away from conventional distillation to the application of gentler solvents (Holmes 1989: 70–3; Kim 2003: 79–83).

The chemist Herman Boerhaave, author of the influential textbook *Elementa chemiae* (1732), approached analysis pragmatically, mixing dry and wet methods as needed. Introducing Daniel Fahrenheit's thermometer into his apparatus enabled him to control the heat with greater accuracy, but he remained skeptical about the possibility of producing true elements through chemical operations. Claude Joseph Geoffroy was the only chemist at the French Academy who still systematically investigated plant chemistry in the 1730s and 1740s, but his studies failed to produce any significant findings. Andreas Marggraf was one of the most successful mid-century analytical chemists. In plant chemistry, his work expanded the scope of solvent extractions while restricting the use of distillation. In 1747, he showed that sugar, such as that produced from sugar cane, existed in other plants, especially beets and carrots. He formulated a method of extracting it from beets using alcohol. Although Marggraf pointed out the economic importance of the discovery, its technical exploitation had to wait until the end of the century (Partington 1961: 723–9; Holmes 1971: 141–3; Holmes 1989; Powers 2014).

Plant materials had thus been at the center of numerous chemical investigations throughout the eighteenth century. While some of these investigations were carried out in dedicated chemical laboratories, many others took place in artisanal and commercial arenas, especially those related to the apothecary, foodstuffs, and dyestuffs trades (Klein 2008). Plant chemistry was an integral part of chemical education and plants appeared in all major eighteenth-century chemical textbooks, including those by Lémery, Boerhaave, Macquer, and Antoine Baumé's *Chymie expérimentale et raisonnée* (1773). As naturalists began to think about plants as living organized beings, chemists around the middle of the eighteenth century started paying more attention to the specific features of a subgroup of proximate principles of plants, which around 1800 became designated as "organic substances." Plant materials were thus studied both as artisanal objects and as natural bodies and objects of conceptual inquiry (Klein 2005a: 264–7). Hence the longstanding focus on the medicinal qualities of plants – which at the early French Academy had prompted a shift from distillation to solution analysis – also influenced the classificatory propensities of chemists.

Pharmacists and chemists traditionally divided plants into "simple" substances or *simplicia* – a group that encompassed remedies found readily in nature, like wax, balsams, resins, and gums – and "composite" substances, known as *composita* or *preparata*. The latter consisted of chemically prepared medicaments, typically the products of extractions, fermentations, and other chemical operations. The distinction between simple and composite substances also echoed the divide between the natural and the artificial. Whereas plants produced remedies "naturally," chemists produced them "artificially." As we have seen, in the first half of the century plant analysis was closely associated with the study of the ultimate principles of natural bodies (whether of the mineral, vegetal, or animal kind). Chemists carefully distinguished between those philosophically informed principles and material *simplicia*, which had long featured as commodities among apothecaries and in other trades.

In the 1750s, however, an important change occurred in this group's epistemic status that illustrated the dynamic relations between natural history and chemistry's theoretical and artisanal features. In 1742, Guillaume-François Rouelle succeeded Boulduc as public demonstrator at the Jardin du Roi, and became the most influential chemical teacher in France until the late 1760s. The young Lavoisier, among other luminaries, attended his lectures. In the 1750s, Rouelle, Venel, and other French and foreign chemists began to group chemically extracted plant *preparata* together with naturally found *simplicia*. The products of this merger, the "proximate principles," were considered as "natural, compound components of plants." As a result, during the second half of the century chemistry experienced an ongoing movement away from the search for the ultimate principles of bodies, which had not been successfully isolated in a pure form, toward substances that could be separated in chemical analysis (Klein 2005a: 286–9). This transformation was a result of changes in chemical concepts, experimental practice, and broader shifts in natural classifications, botany, and physiology. Until the 1790s, based on the products of various analytical methods – forms of distillation, the use of solvents, and mechanical separations – chemists included mineral substances (especially salts) among the proximate principles of plants. Supported to an extent by Lavoisier's new theory that organic substances were composed from carbon, hydrogen, oxygen, and nitrogen, after the 1790s chemists began to gradually exclude minerals from this class, which in the nineteenth century came to signify "organic" substances (Klein 2005a: 295–323; Klein 2005b; Klein and Lefèvre 2007: 195–253).

MINERAL CHEMISTRY: SALTS, COMPOSITION, AFFINITIES

Whether chemists sought the elementary constituents or the medical essences of substances, chemical analysis was central to their practice. The disputes over analytical methods shaped new notions of chemical elements and composition, which in turn reoriented the aims of experimental chemistry. If Lémery had initially seen chemistry as the art of "separate[ing] the different Substances which are found in mixt Bodies," by 1720 his definition expanded to that of an art that "teaches how to separate the useful parts of a Body from the unuseful, and how to join them together again" (Lémery 1720: 1). The emergence of new operative attitudes to chemical elements and composition demonstrates the growing importance of chemical *synthesis* as an experimental counterpart to *analysis*. Cycles of successful analysis and re-synthesis of substances confirm chemists' concern with operations. As the century advanced, chemists turned their attention to mapping out the selective tendencies of an increasing number of substances to combine.

Homberg described chemistry in 1702 as "the art of reducing compounds into their principles by means of fire, and of composing new substances in the fire by the mixture of different materials" (Eklund 1975: 2). Boerhaave highlighted the importance of synthesis as a criterion of proper analysis, specifying that "by our Art Bodies are separated … into their natural constituent parts" that "only shou'd be extracted, which being afterwards properly compounded together, we are certain, would again produce exactly the very same body" (1735: vol. II, 2). According to Macquer, the "chief End of Chymistry is to separate the different substances that enter into the composition of bodies; to examine each of them apart; to discover their properties and relations; to decompose those very substances, if possible … [and then] to reunite them again into one body so as to reproduce the original compound with all its properties" (Macquer 1758: vol. I, 1).

The shift in plant chemistry towards solution analysis enfeebled the status of distillation and its attendant five principles. At the same time, it cast a long shadow on mineral chemistry and especially the study of salts and metals. Following the old maxim "like attracts like," chemists had traditionally assumed that substances combine due to a correspondence or sympathy between them. The mixing of acids and alkalis, which produce salts, presented an intriguing exception. Contrary to previous views of combination, chemists interpreted acid–alkali reactions as processes of mutual destruction. Ever attentive to qualitative changes, what we call neutralization, they depicted using words like strife, nullification, and combat. Acids and alkalis, in this view, reacted in a consistently marked way not because of shared similarities but because they were antagonistic. The distinctive sensory features accompanying such reactions, above all effervescence and ebullition, came to define the acid or alkaline quality of a body. Acids and alkalis were defined in relation to one another. If a substance effervesced with an alkali, it was an acid, or if it effervesced with an acid, it was an alkali. Boyle's pioneering work on color indicators, which was further developed in the eighteenth century, enabled chemists to reliably identify both acids and alkalis. Acids turned all vegetable indicators (like syrup of violets, oak galls, and litmus) red; alkalis turned them green; and neutral substances produced no color change. Throughout the eighteenth century, the acid–alkali reaction was one of the most familiar laboratory experiences (Siegfried 2002: 74–9; Kim 2003: 111–12).

Salts, the products of acid–alkali reactions, can be acidic, alkaline, or neutral. The concept of a neutral salt, which was a result of experimental practice, offered chemists a new way of understanding and controlling chemical composition. Here was an experimental setup in which reactants and products as well as the reaction itself could be reliably identified. In the case of neutral salts, the product was neither acidic nor alkaline. It bore no qualitative resemblance to the two substances from which it was prepared, and the distinctive qualities of the components could not be identified in the neutral compound. As we have seen, analogous difficulties plagued analytical methods in plant chemistry, but whereas chemists had no empirical way of ascertaining the pre-existence of the products in the materials they subjected to distillation or solution analysis, it was experimentally evident that salts, including neutral ones, were composed of acids and alkalis. Moreover, through chemical manipulation these components could be regenerated from solutions of the salts. The focus on the acid–base–salt relationship exemplified chemists' growing interest in operationally based notions of composition and simple bodies, at the expense of their previous concerns with elements in the ultimate meaning of the term. Such practical and conceptual adjustments would eventually contribute to Lavoisier's operational redefinition of a chemical element as the endpoint of chemical analysis (Siegfried and Dobbs 1968). The experimental and theoretical priorities of eighteenth-century chemists steadily evolved: from ultimate to proximate principles; from metaphysical elements to operationally determined simple bodies; and from the general-abstract attributes of matter to tangible substances and their particular behaviors.

Chemists realized that metals dissolve selectively in acids and that alkaline salts precipitated them from such solutions. Lémery wondered why the solvents "quit the Bodies they held before in Dissolution, to betake themselves to some other … why the *Aqua Regalis* [mixture of nitric and hydrochloric acids] leaves the Gold it was impregnated with, to give way to the Alkali Salt." He offered a mechanistic explanation, yet noted that "this question is one of the most difficult to resolve well, of any in Natural Philosophy" (Lémery 1677: 26; Lemery 1720: 67). E.F. Geoffroy approached solution chemistry from a more operational standpoint. Instead of dwelling on the possible causes and mechanisms underlying displacement reactions, he tabulated the order of these dispositions to combine. The result, the first table of chemical affinities, appeared in 1718 as "Table of different rapports observed in chemistry among different substances." Other chemists soon emulated this model. On one count, the period 1730–1800 saw the publication of affinity tables by at least two dozen different authors.

Geoffroy's own table consisted of sixteen columns. At the head of each column was a reference substance, followed by a list of substances arranged according to the order in which they would replace one another in combination with it (Figure 2.2). Scholars have debated the institutional, practical, and epistemic

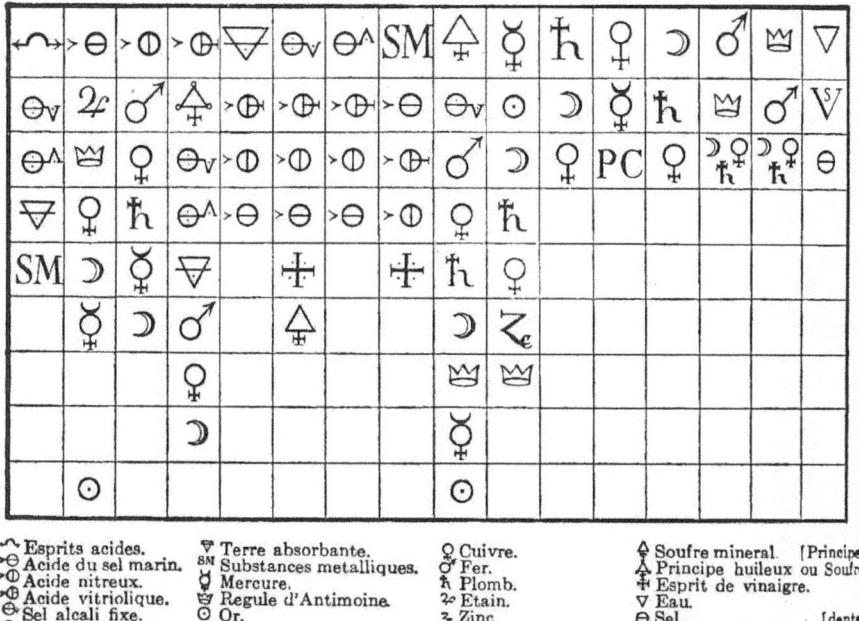

FIGURE 2.2 Geoffroy's 1718 affinity table. From E.F. Geoffroy, "Table des differents rapports observés entre differentes substances," *Mémoires de l'Académie Royale des Sciences* (Paris: Imprimerie Royale, 1718), p. 212, plate 8. Sourced from Wikipedia.

dimensions of Geoffroy's table. By using "rapport" instead of "attraction," a term that might have implied the action of Newtonian forces, Geoffroy underscored the relationships among the substances and their empirically discernible tendencies to combine, rather than speculating on the nature of the causes underlying them. As more substances and reactions became known such tables proliferated, adding a predictive dimension to chemical practice. One of the most iconic instances was compiled by the Swedish chemist Torbern Bergman, who first published it in 1775, followed by an enlarged and revised version in 1783. Bergman's elaborate table was the first to explicitly show the change in the order of affinities due to heat as well as other exceptions. It was accompanied by a detailed Dissertation on Elective Attractions, comprising fifty-nine columns, and referred separately to both dry and wet analytical methods (Roberts 1991b; Klein 1995; Holmes 1996; Duncan 1996).

Marggraf, the discoverer of beet sugar, also made significant contributions to mineral chemistry. He used precipitation methods for analysis, such as the Prussian blue reaction for the detection of iron, and improved the process for obtaining phosphorus from urine. In 1746, he isolated zinc by heating calamine (a zinc ore) and carbon. Marggraf was known for his sparing use of hypothesis and great operational skills, two hallmarks of an accomplished experimental chemist. His work was characterized by the use of quite small amounts of materials. While many of his contemporaries worked with large samples, sometimes weighing pounds, thanks to his advanced operational skills Marggraf was able to obtain accurate results often using less than an ounce (Partington 1961: 724). He also used a microscope to inspect the appearance of minerals and crystals.

For mineral analysis Marggraf employed blowpipes. A remarkably simple instrument used by glassblowers and metalworkers for millennia, the blowpipe typically consisted of a narrow brass or silver tube through which a stream of air was blown either by mouth or bellows. By directing the air jet through a flame onto a small test sample the size of a peppercorn, very high temperatures could be attained. Crucial to the successful use of the blowpipe was circular breathing, the ability to inhale through the nostrils while continuously expelling air from the mouth. In this way samples, often placed on small blocks of charcoal, could be continuously heated for extended periods in the presence of other reagents, and the experimenter could carefully observe their thermal behavior, monitoring flame coloration, sublimation, decomposition, etc. First used by German chemists around the middle of the eighteenth century, the blowpipe later gained popularity among Swedish mineralogists and metallurgists. It was instrumental in identifying metallic elements and the qualitative composition of most minerals until the middle of the nineteenth century. In his book on the blowpipe, *De tubo feruminatorio* (1779), Bergman extolled its virtues as a simple, portable, and rapid means of qualitative although not quantitative analysis (an important distinction). A blowpipe virtuoso, the Swedish chemist Jacob Berzelius published in 1820 the first treatise on the use of the blowpipe that was influential in France and England (Szabadváry 1966: 50–5; Jensen 1986; Niinistö 1990; Dolan 1998; Abney Salomon 2019).

PNEUMATIC CHEMISTRY: AIRS FROM HALES TO PRIESTLEY

Pneumatic chemistry, or the chemistry of gases, was one of the most rapidly evolving branches of eighteenth-century experimental science. Momentous developments in this area, like Lavoisier's oxygen theory of combustion and the establishment of the chemical properties of distinct gases, were key components of the chemical revolution of the 1770s and 1780s. As one of the most gifted and prolific pneumatic experimenters of his generation, Priestley played a prominent role in this scientific story. In 1776, he enthusiastically announced that pneumatic chemistry was "not now a business of *air* only ... but appears to be of much greater magnitude and extent, so as to diffuse light upon the most *general principles* of natural knowledge, and especially those about which *chymistry* is particularly conversant." "By working in a tub of water, or a basin of quicksilver," he boasted, "we may perhaps discover principles of more extensive influence than even that of *gravity* itself." As late as 1790, by which time the revolution was all but over, Priestley still echoed his earlier experimental enthusiasm (1775–1777: vol. II, vii–viii; Priestley 1790: xxiv). Priestley's statements present a telling contrast between the lofty aspirations of pneumatic chemists and the simplicity of their apparatus. Pre-revolutionary pneumatic research depended in large measure on the practitioner's skill, experience, and ingenuity.

The Flemish physician and chemist Jan Baptist van Helmont is credited with the first use of "gas" to denote what were then usually called "airs." "I call this Spirit, unknown hitherto," he explained, "by the new name of Gas, which can neither be constrained by Vessels, nor reduced into a visible body." Far removed from modern notions of gas, Van Helmont's definition betrays the chief concerns of contemporary chemists. Gases or airs were not only empirically elusive in this period, but also dangerous to work with. Boyle had observed that "not finding the Air to be a visible Body," most experimenters ignore it, for "what is invisible, they think to be next Degree to nothing." Several decades later, the Newtonian experimenter Stephen Hales wrote of a "much neglected volatile *Hermes*, who has so often escaped through … burst receivers, in a guise of a subtle spirit, a mere flatulent explosive matter." Experimenting in the 1660s with an air pump, Boyle discovered the inverse relation between the pressure and volume of air, known today as Boyle's law. But to the pneumatic chemists this aerial pressure raised much more immediate concerns. Its build-up in closed systems, such as distillation apparatus, often caused vessels to explode, ruining hardware and endangering the operators. To avoid this kind of experimental havoc, chemists made holes in their apparatus. The problem of handling expanding vapors during distillation and preventing them from escaping was sometimes addressed by placing oversized vessels at the end of the spout on the distilling head. These vessels provided extra volume and enabled greater condensation. They also helped correct the common problem of heat control (Eklund 1975: 11; Crosland 2000: 80–2; Levere 2001: 51–2).

The subtlety, expansibility, and volatility of gases presented considerable theoretical and practical challenges, many of which remained unanswered for most of the eighteenth century. The quantitative examination of the volume and elasticity of airs offered little to chemists, for it did not contribute to a qualitative understanding of the chemical nature of gases. Boyle, who maintained a lifelong interest in air and its qualities, observed that a candle could not burn in an airless space. Yet even as chemists became increasingly aware of the atmosphere's role in combustion, respiration, and other phenomena, they did not attribute this to the chemical properties of air but to substances that the air – primarily conceived as the medium surrounding us – could absorb and release (Levere 2001: 51–2). Boyle suspected that air was not "a Simple and Elementary Body, but a confus'd Aggregate of Effluviums" and possibly the most "heterogeneous

Body in the world." Concluding his landmark analysis of "Air by a great variety of chymio-statical Experiments," Hales conjectured that "our atmosphere is a *Chaos*, consisting not only of elastick, but also of unelastick air-particles, which in great plenty float in it" (Boyle 1674: 2; Hales 1727: 155, 315).

The origins of pneumatic chemistry are often traced to Hales' 1727 *Vegetable Staticks*, in which he suggested that air should be considered a ubiquitous and powerful chemical agent. Although "it has hitherto been overlooked and rejected by Chymists," he reflected, "may we not with good reason adopt this now fixt, now volatile *Proteus* among the chymical principles?" (Hales 1727: 316). Hales' experimental program was based on a modest device – the pneumatic trough – whose widespread use revolutionized the practical, and ultimately theoretical, study of airs (Figure 2.3). Priestley alluded to this apparatus when he referred to the groundbreaking contributions accomplished by working in a tub of water or mercury. Characteristically humble and generous, Priestley added: "my apparatus for experiments on air is, in fact, nothing more than the apparatus of Dr. Hales, Dr. Brownrigg, and Mr. Cavendish, diversified, and made a little more simple" (1775–1777: vol. I, 6). What set the pneumatic trough apart from previous attempts to capture airs was due to the way it separated the generating from the receiving vessel. Before Hales, experimenters like Boyle and John Mayow had developed techniques to generate and trap airs in cupping glasses inverted over water. They occasionally used burning lens to initiate the chemical reaction within the glass cup, and they acquired the dexterity to collect the newly produced air by "pouring" it under water (Parascandola and Ihde 1969: 352–3).

Because Hales did not distinguish between different kinds of air, it would be wrong to assume that he invented the pneumatic trough for the purpose of isolating airs in their pure state. Air and that chaotic atmosphere, also known

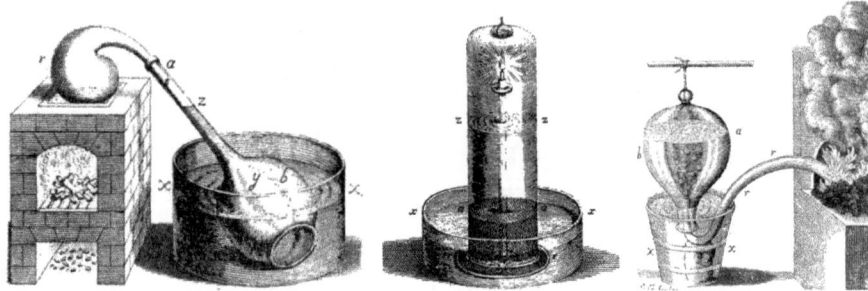

FIGURE 2.3 Hales' two-vessel apparatus for measuring amounts of gas produced or absorbed; pedestal apparatus; iconic pneumatic trough. From S. Hales, *Statical Essays: Containing Vegetable Staticks; Or, An Account of some Statical Experiments … Also, a Specimen of An Attempt to Analyse the Air, by a great Variety of Chymio-Statical Experiments*, 3rd ed. (London: Innys and Manby, 1738), vol. 1, pp. 168, 211, and 266. Courtesy of HathiTrust.

as common air, were for Hales one and the same thing. He therefore made no systematic attempt to study the properties of the gases he was so efficiently able to isolate and collect. Instead, he carefully measured the volume of air that could be generated from and absorbed by various substances during heating, solution, and fermentation. Hales' distinction between "elastick" and "unelastick air-particles" corresponded to his view of air as being "now fixt, now volatile." When fixed within bodies, i.e. in the solid state, air assumed an inelastic form; when it was freed from combination, it regained its natural elasticity (Parascandola and Ihde 1969; Levere 2001).

Hales described several similar devices for the isolation and collection of airs. A glass vessel was filled with water, inverted, and placed into a shallow water-filled trough. A gas-generating retort was then connected to the inverted vessel so that gas could bubble up through it, gradually displacing the water it contained. The same effect could be accomplished using a single vessel, partly filled with water and inverted over a water trough. A pedestal or a platform placed inside the vessel, above the water level, created the space for the materials and reaction to take place. In this one-vessel apparatus the airs would partly dissolve in the water above which they were generated and collected (Figure 2.3). The diminution in aerial volume, due to the air's loss of its soluble components, could be avoided in models in which the generator was separate from the collector. Some experimenters used mercury instead of water. The pneumatic trough gained initial prominence because it allowed operators to "wash" airs of water-soluble impurities and enabled their collection and storage in glass receivers (Parascandola and Ihde 1969: 353–7; Crosland 2000).

Hales' pneumatic trough became more important in the hands of the chemist and physician Joseph Black, even though Black did not use the apparatus in his seminal work on magnesia alba (magnesium carbonate). Black had initially sought a cure for bladder stones, which he knew could be dissolved in caustic alkali. Because caustic alkali could not be safely ingested, he turned his attention to a much milder alkali, magnesia alba. Although his search for an oral medicine for the stone failed, Black discovered that when heated, magnesia alba yielded an air. Unlike common air, when passed through a clear solution of lime in water this air turned the limewater milky, which meant that small solid particles were being created in the solution. Black encountered the same air when he treated limestone with mineral acids, and when heated limestone turned into quicklime. He concluded that there were two airs, or two kinds of air: atmospheric air and the air that had been held in combination in solid magnesia alba or in limestone, which he called "fixed air." Black showed that the latter kind, which we recognize as carbon dioxide, was also produced in respiration, fermentation, and combustion. He carried out a cycle of quantitative experiments in which a balance was used at all stages, and he posited that the weight lost during heating was due to the gas generated without actually collecting it. Chemists, metallurgists, and pharmacists routinely used weighing techniques in their practice, and had done so for a long time, but Black was the first to apply gravimetric principles to the study of airs. Using a balance that was accurate to at least one part in 200, he showed that fixed air was not a version of atmospheric air, but was a different and distinguishable air (Guerlac 1957a; Guerlac 1957b; Levere 2001: 54–5).

Black was the first chemist to show that gases could be chemical substances in themselves and not just atmospheric air in different states of purity. Because airs, including fixed air, were so light, Black had to measure their weight indirectly, by determining the weight loss suffered by the heated magnesia alba or limestone. Black's method resulted not only in the discovery of fixed air as the first aerial species distinct from common air, but also served as the prototype for the quantitative methods later generalized by Lavoisier.

Henry Cavendish pursued similar quantitative analyses, employing a combination of pneumatic instruments similar to the ones designed by Hales, outstanding experimental skill, and attention to details and weight measurements (Figure 2.4). In 1766, Cavendish announced the discovery of "inflammable air," our hydrogen. Igniting it in the presence of atmospheric air, he detected a contraction in the volume of the gases and proceeded to weigh the newly formed droplets of water. This observation provided him with early insights concerning the composition of water. In the 1770s and early 1780s, employing modest, partly improvised equipment – similar, once again, to the apparatus used by Hales, Black, and Cavendish – Priestley rediscovered the airs isolated by his predecessors while adding an impressive list of his own (Figure 2.5). The list included dephlogisticated air (oxygen), nitrous air (nitric oxide), alkaline air (ammonia), marine acid air (hydrochloric acid), and phlogisticated

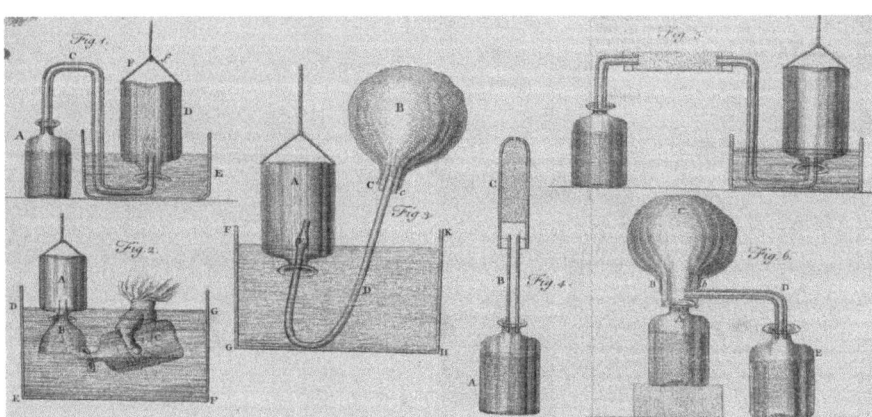

FIGURE 2.4 Cavendish's pneumatic instruments and method. From H. Cavendish, "Three Papers, Containing Experiments on Factitious Air," *Philosophical Transactions of the Royal Society* 56 (1766), table VII, p. 141. Courtesy of Royal Society Publishing.

FIGURE 2.5 Priestley's pneumatic instruments and workshop. From J. Priestley, *Experiments and Observations on Different Kinds of Air, and other branches of natural philosophy ... Being the former six volumes abridged and methodized, with many additions* (Birmingham: Pearson and Johnson, 1790), vol. 1, plates I and II. Courtesy of HathiTrust.

air (nitrogen). Combining practical and material ingenuity with a tireless inquisitive spirit, Priestley was the most prolific pneumatic experimentalist of his era (Schofield 1997; Schofield 2004; Boantza 2013a: 145–70).

REVOLUTIONARY CHEMISTRY: LAVOISIER'S NEW EXPERIMENTAL ORDER

Led by Lavoisier in collaboration with Louis-Bernard Guyton de Morveau, Claude-Louis Berthollet, Antoine François Fourcroy, Pierre-Simon Laplace, and others, the chemical revolution was a complex affair that transformed chemical theory, language, and practice. Lavoisier is the most fitting representative of the new chemistry. One of the hallmarks of Lavoisier's new chemistry was its overall conceptual and experimental integrity. The novel standardized nomenclature and notation represented new notions of chemical composition, element, reaction, and classification that were in turn demonstrated and confirmed by innovative instruments and investigative methods. The bulk of Lavoisier's experimental innovations stemmed from his conscientious approach to the principle of weight conservation in chemical reactions. In practice, this meant the subjection of chemical matter and phenomena to precise gravimetric measurement and calculation. As Lavoisier stated in his *Elements of Chemistry*, "in all the operations of art and nature, nothing is created; an equal quantity of matter exists both before and after the experiments ... Upon this principle the whole art of performing chemical experiments depends" (A.-L. Lavoisier 1790: 130–1). The analytical balance became the physical and metaphorical embodiment of the new chemistry; it was the prime instrument in Lavoisier's overarching quantification of chemistry. Lavoisier's most sensitive balance could weigh to an accuracy of 1 part in 400,000. As this astonishing level of precision far exceeded the purity of his samples and reagents, it was experimentally unwarranted, but the results could be carried to many significant figures, sometimes to great rhetorical effect (Stock 1969; Levere 1990; Golinski 1995).

Chemists of differing backgrounds – including metallurgists, mineralogists, and pharmacists – had long used balances to carefully weigh substances, but their mode of chemical reasoning and practice was predominantly qualitative. Seventeenth-century chemists like Van Helmont, Starkey, and Boyle had been aware of the importance of weighing and measuring materials before and after experiments, and so did Black and Cavendish, as we have seen. Beyond practical applications, however, many chemists in the middle of the eighteenth century still associated weight with physical inquiries. Physicists studied the nature of bodies by analyzing measurable attributes like their size, shape, weight, and motion. For many pre-revolutionary chemists, a body's weight carried no more scientific import than, for instance, its appearance, taste, solubility, or tendency to react with other substances in specific ways under particular conditions. Generally speaking, before Lavoisier chemists weighed for practical, not theoretical reasons. After Lavoisier, chemistry was conceptually and experimentally directed by gravimetric considerations far more than previously. Lavoisier did not forgo the qualitative aspects of chemical practice, many of which still functioned as standards for the identification of various materials and reactions, but he turned the conservation of weight in chemical reactions into a governing principle in chemistry.

This meant that *all* matter, reactants and products alike, had to be accounted for by weight, either directly or indirectly. Elusive and protean entities, from very light gases to "imponderable fluids" like heat (and possibly light), had to be included in the grand scheme of quantification. This entailed practical challenges, some of which took Lavoisier well over a decade to work out. He benefitted from the assistance of his highly capable colleagues and his wife, Marie-Anne Paulze, as well as eminent instrument makers like Jean Nicolas Fortin and Pierre Mégnié (Beretta 2014b). The prevalent theory of combustion before Lavoisier was the phlogiston theory. Its core assumption, anchored in traditional, qualitative notions of chemical change, was that when a body underwent combustion or calcination, a certain presumed constitutive principle of inflammability, phlogiston, flew away. It made eminent qualitative sense to conceive of a burned substance or the crumbly remains of a roasted metal as having lost something, and it stood to reason that the emission of heat and light during deflagration was the result of the release of such a principle.

Many chemists before Lavoisier had observed that metallic calxes weigh more than the original metal. With the growing importance of weight in chemical reasoning, it became increasingly difficult to explain this phenomenon in terms of phlogiston. In the early 1770s, the Dijon lawyer and chemist Louis-Bernard Guyton de Morveau discovered through a series of careful quantitative experiments that the relative weight gain for each calcined metal was fixed. This prompted Lavoisier to think of calcination as a process not of loss of phlogiston but of fixation of something from the air. In 1774, during a visit to Paris, Priestley told Lavoisier about the new (dephlogisticated) air he had obtained from the heating of mercuric oxide with a burning lens. Applying his new gravimetric principles and methods, Lavoisier repeated the experiment and noted that this new air, which had been released by a calx, was qualitatively different from the air that Black had discovered, in that it supported combustion and respiration. Yet it could combine with charcoal to form Black's fixed air. Lavoisier's conclusion was that the new air – which he called oxygen and (incorrectly) identified as an acid-generating principle – was a constituent portion of the atmosphere (Levere 2001: 63–4). Lavoisier thus laid the foundations for a new theory of combustion and respiration, and paved the way to determining the constitution of the atmosphere as an aerial mixture of chemically distinct gases. Priestley's dephlogisticated air became Lavoisier's oxygen.

From around 1772, for nearly two decades Lavoisier focused on processes involving the fixation and release of airs, including combustion, calcination, fermentation, plant growth, and respiration. He began his career as an experimental chemist using relatively simple and traditional instruments and experimental setups, much like the ones used by Hales and Priestley. But by the 1780s, he had assembled an extensive array of expensive apparatus, exclusively designed and crafted for him (Figure 2.6). Alongside his unequalled state-of-the-art balances, he used gasometers and a calorimeter. Lavoisier's design and use of these instruments illustrates both the central challenges he was facing and the ways he devised to overcome them. The gasometer, which went through several iterations during the 1780s, allowed him to accurately handle gases and their flow. The ice calorimeter, which emerged out of his collaboration with the physicist Laplace, enabled him to gravimetrically account for the heat loss in various chemical and physiological reactions. Although heat could not be directly measured by weight, Lavoisier's calorimeter enabled the experimenter to account for the weight of the water produced by the melting of the ice surrounding the chamber in which the reaction took place. Thus, the quantity of heat produced in a chemical reaction could be measured indirectly, through its effects, while its cause – be it the presence of a unique matter of heat or the motion of ordinary matter – remained undetermined (Holmes 1988; Holmes 1997). As we have seen, the success of many chemical experimental schemes crucially depended on the practitioner's skill, the capabilities of the instruments involved, or a combination thereof. Lavoisier was a highly competent experimenter, but some of the instruments he employed during the chemical revolution represented a major break with the long tradition of

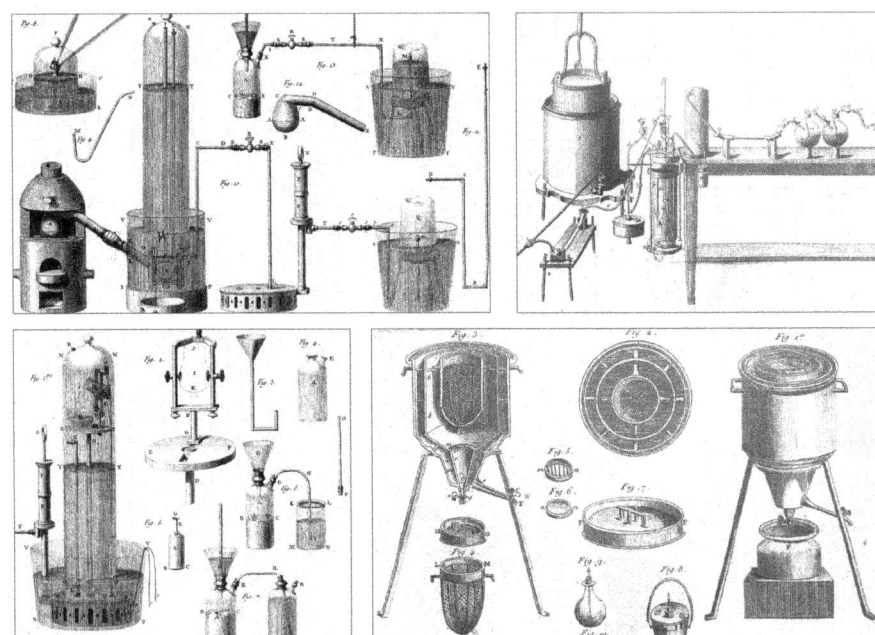

FIGURE 2.6 Examples of Lavoisier's instruments from the early 1770s (left) and late 1780s (right), including his ice calorimeter (bottom right) for quantifying heat in chemical reactions. A.-L. Lavoisier, *Opuscules physiques et chymiques* (Paris: Durand, 1774), vol. 1, plates I and II; A.-L. Lavoisier, *Elements of Chemistry, in a New Systematic Order, Containing All the Modern Discoveries*, trans. R. Kerr (Edinburgh: Creech, 1790), plate XI. Courtesy of HathiTrust.

embodied skills and experiential knowledge on which generations of chemists had prided themselves. These instruments enabled Lavoisier to both reveal previously imperceptible changes and to implement quantification standards at an unrivaled degree of accuracy[3] (Levere 1994: iv).

One of the best-known experiments Lavoisier performed was the demonstration of the composition of water, which was inspired by Cavendish's observations. Using an apparatus that allowed him to control the flow of two gases, in 1783 Lavoisier conducted a continuous combustion of hydrogen (inflammable air) and oxygen. To conclusively prove that water was the *only* product of this reaction, he employed his characteristic gravimetric method, for which he ultimately needed to develop the gasometer. With input from the mathematician Jean Baptiste Meusnier and Mégnié's craftsmanship, by the end of 1785 Lavoisier had two rudimentary gasometers at his disposal. Able to control the flow of gases by volume, he could continuously burn hydrogen with oxygen in a two-to-one ratio by volume. Knowing the pressure and temperature of the gases enabled him to calculate their weight and thus to argue gravimetrically and convincingly that water was the sole product of the combination of hydrogen and oxygen. However, Lavoisier had the means, ambition, and institutional support to push precision even further. For public demonstrations he used his precision gasometer. Evocatively portrayed by a modern commentator as "the Rolls Royce of chemical instruments" and labeled the "fruit of a marriage between Industrial Revolution engineering and physics," this was Lavoisier's most impressive and luxurious instrument. Two models were presented to the Academy in 1788 (Levere 2001: 70–4; Levere 2005; Chang 2012).

Similar sensibilities and commitments can be seen in Lavoisier's other experimental programs, including those on respiration, heat, and organic chemistry (Guerlac 1976; Holmes 1984; Beretta 2005). From a chemical perspective, for the first time the three realms of nature had been all but materially unified. Water was not an element anymore; earths, partly designated by salts, were now composed of acids and bases[4]; fire had become the subtle and imponderable matter of heat, caloric; and air had now turned into a mixture of distinct gases like oxygen, hydrogen, and nitrogen, which represented both chemical species and a universal–physical state of matter.

In 1802, in an introductory note to his series of "Lectures on Chemistry" at the Royal Institution in London, Humphry Davy explained the relevance of chemistry to all areas of science, from physics to physiology and from mineralogy to botany. Chemistry was not merely an applied form of inquiry or a particular craft subordinated to other scientific disciplines. Davy thus portrayed the chemist as someone who "science ... has bestowed upon him powers which may be almost called creative; which have enabled him to modify and change the beings surrounding him, and by his experiments to interrogate nature with power ... as a master, active with his own instruments." Like many chemists before him, including the ones we have encountered, as well as many after him (like Marcellin Berthelot), Davy emphasized the creative powers of practicing chemists, anchored in experiential knowledge and the skillful handling of substances, reactions, and instruments. Practice and experiment have always been at the core of chemistry, which early in the eighteenth century was still partly subordinated to alchemical, metallurgical, and medical goals. As the century wore on, distillation and solution methods, originally used for the procurement of medicaments, turned into analytical tools closely related to new notions of composition and chemical elements. Time-honored accurate weighing techniques, always crucial to practicing pharmacists and metallurgists, became in the hands of Lavoisier the embodiment of a new chemical theory (Davy 1839: 319).

Chemistry's role in the production of useful materials and its longstanding ties to industrial pursuits have changed over time, but never faded. During the eighteenth century, however, chemistry had gained a significant measure of disciplinary and professional independence. For Davy, one of the most skillful chemical experimenters and demonstrators of the early nineteenth century, chemistry was foundational. Still profoundly rooted in experiment, practice, and material creation, it had become a science that "has for its objects all the substances found upon our globe ... The phenomena of combustion, of the solution of different substances in water, of the agencies of fire ... and the conversion of dead matter into living matter by vegetable organs, all belong to chemistry" (Davy 1839: 311).

CHAPTER THREE

Laboratories and Technology

MARCO BERETTA

INTRODUCTION

The eighteenth century was an exciting period in the history of chemistry. The number of newly discovered metals, earths, alkalis, and gases was greater than in all the preceding history of chemical science, and contributed to the collapse of older philosophies of matter (Weeks 1933).[1] Important as they were, these discoveries were not the result of new theoretical ideas; rather, they reflected the increasing accuracy of analytical methods and apparatus. Such technical progress was made both in the laboratory and in the field. The new gases, for instance, were both isolated in laboratories and (in the case of methane) discovered in marshes. The different circumstances in the discoveries of new elements implied the existence of a wide range of types of apparatus, from the portable pocket laboratory (Smeaton 1966), to the pursuit of chemical experimentation on an industrial scale.

EARLY LABORATORIES AND THE CHEMICAL CRAFTS

Because of the variety of actors and institutions involved in the development of chemical laboratories and technologies during the eighteenth century, its history is complex. Sites, techniques, and materials changed, sometimes substantially, following the different array of needs in the various chemical specialties and crafts. Although fire dominated chemical practice, furnaces, tools, combustibles, reagents, and spaces differed from trade to trade. Glassmaking, for instance, required the use of particular furnaces, tools, combustibles, and reagents that were hardly comparable with those in use in pharmacy, dyeing, tanning, assaying, mining, smelting, or distilling.

To complicate the picture further, chemical trades were regulated and administered in profoundly different ways. Large glassworks such as Saint Gobain near Paris were often capitalistic enterprises in which the division of labor and the technology in use were the products of a policy that was relatively independent. By contrast, smaller workshops, where most of the common chemical glassware was produced, were run by artisans who belonged to municipal guilds that were subjected to strict regulation and control (Beretta 2012a). Pharmacies were also regulated by the statutes of their guild, and their scientific output remained under the control, not without conflict, of medical faculties until the second half of the eighteenth century. Because of the need for large investments, mining works were, for instance in Sweden, regulated and supported by state agencies.

During the eighteenth century, despite the complexity of the overall picture, the chemical trades made important progress, and many able artisans distinguished themselves with inventions and ideas that gave chemistry an unprecedented prestige. It is therefore not surprising that it was precisely in this period that academic positions in chemistry were created throughout Europe in both universities and scientific academies. Behind this process of academic institutionalization there was an implicit expectation that chemistry would become a science, and, through the development of new theories, could eventually be able to coordinate all the chemical arts, techniques, and practices, and clarify the principles behind experimentation. This process also explains the publication of instructions on how to build a laboratory and apparatus, such as Peter Shaw and Frances Hauksbee's *An Essay for Introducing a Portable Laboratory* (1731), and Robert Dossie's *The Elaboratory Laid Open or, The Secrets of Modern Chemistry and Pharmacy Revealed* (1758) (Eklund 1975).

In time, academic chemists also designed their own laboratories that enabled the opening of new important research fields. However, the creation of university positions and laboratories could never substitute for the sites and practices created within the chemical trades. In fact, artisanal activities, especially in pharmacy and mining, had a strong influence on the development of eighteenth-century academic chemistry. The creation of the chair in chemistry at Uppsala University in Sweden was principally guided by the state's need to expand the number of academic chemists who could work in mining ventures, one of the main assets of the Swedish economy (Fors 2015).

Academic chemists were able to influence artisanal chemistry in other ways. In Scotland and the Netherlands the creation of new chairs in chemistry was intimately connected with the need to provide medical faculties with professional and academic pharmaceutical chemists. Joseph Black's investigation at the University of Edinburgh of the chemical properties of gases (1754) began in a medical context, but it opened an entirely new path of experimental investigation that was destined to play an extremely important role within chemistry more broadly. Moreover, Black played an important role in the development of several chemical manufacturers.

Even in the first decades of the century academic chemists began to have a clear idea that their science was neither entirely philosophical like mathematics or astronomy, nor entirely practical as many people still believed it to be. In a draft devoted to the usefulness of applied sciences presented to the Académie Royale des Sciences of Paris between 1716 and 1727, one of the most authoritative French scientists, René-Antoine Ferchault de Réaumur, pointed out the central role played by chemistry in the growth of several trades in the following terms:

> Chemistry, whose investigations seem rather frivolous to those who do not know its true purpose, could become one of the most useful parts of the Academy. ... The conversion of iron into steel, the method of plating or whitening iron to make tin-plate, the conversion of copper into brass, three great industries which the Kingdom lacks, are in the province of chemistry. It is the business of chemistry, also, to investigate the mineral substances used in dyeing and ores and minerals. Glassworks, pottery works, faïencers, porcelain – industries which all need to be improved – also concern it.
>
> (Réaumur 1716–27 [1888])

A few years later, Herman Boerhaave, professor of medicine and chemistry at the University of Leiden, also acknowledged the proximity of chemistry to the chemical arts, and proclaimed his confidence that by perfecting the instruments and apparatus used in laboratories, the science could be radically transformed (Powers 2012).

Local traditions also played a very important role in fostering specific arts and practices. Since the late Renaissance, Sweden and several German states had relied on solid metallurgical and mineralogical bodies of knowledge. In the Netherlands, Italy, and Scotland, thanks to the chairs created in the first half of the century, chemistry became an academic field that progressively helped to transform medical practice and, at the same time, to create a large number of educated academic pharmacists. In France and Germany the prestige of the Académie Royale des Sciences and of the Prussian Academy of Sciences helped chemists to enhance their social prestige (Klein 2009).

In Britain, a growing collaboration between academic chemists, artisans, and entrepreneurs played a key role in transforming the scale of chemical practice, which partly underpinned the industrial revolution (Clow and Clow 1952). Thanks to the growing efforts to explain the discovery of new elements, the chemical properties of gases, and the theory of latent heat, with the help of more sophisticated theories, the gulf between academic and artisanal chemistry narrowed during the second half of the century, hence diminishing the differences between sites and apparatus.

At the beginning of the eighteenth century ordinary laboratories consisted of one or two rooms, but by the end of the century their dimensions and architectural design had expanded considerably both in size and design. Similarly, the tools used by chemists at the beginning of the century were nearly identical to those used during the Renaissance, but with the discovery of gases and especially during the 1780s there was substantial innovation in chemical practice resulting in a a large number of new and complex apparatus.

LABORATORIES AND MATERIAL CULTURE BEFORE THE DISCOVERY OF GASES

The facilities in which chemists of the eighteenth century operated showed marked differences, for example, between the site of public demonstrations used for teaching purposes and the private laboratory where the most capable students were trained to do sophisticated experiments and, at times, even research. Although important innovations in apparatus were made in several chemical specialties, principally in mineralogical chemistry, it was pharmacy which had the greatest influence on the features of the eighteenth-century chemical laboratory.

There were several reasons for the importance of pharmacy for the evolution of the chemical laboratory. Above all, pharmacists were by far the largest community of artisanal scientists practicing chemistry in Europe. In 1756 there were 130 apothecary shops in Paris alone (Bouvet 1937: 264). Moreover, laboratory practice was a daily routine for apothecaries. Their work included the manipulation of chemicals and the accurate preparation of remedies, as well as their accurate weighing, thereby contributing to setting the standards for apparatus, weights, and measures. In addition to running their private laboratories and shops, pharmacists taught chemical courses in medical faculties, botanical gardens, and other academic establishments. In these institutions apothecaries introduced large audiences to relatively simple chemical operations and made chemistry an extremely popular science. Last but not least, the social relevance of pharmacy is underlined by the remarkable fact that out of 120 apothecary shops that were active in Paris in 1776, twenty-two were run by widows of deceased pharmacists (*État* 1776: 138–46). Such a large female presence was a consequence of the role played by women in the history of ancient and Renaissance alchemy, as well as their long-established skills in the preparation of cosmetics and perfumes.

Chemistry became so fashionable in Paris in the second half of the eighteenth century that the yearly attendance in the courses given in the French capital reached 3,130 students in 1781 (Lehman 2008; Perkins 2010). Although significant progress emerged in other European cities, especially in Germany (Hufbauer 1982; Klein 2007), Parisian chemists appropriated the innovations very rapidly and quickly adapted their courses and the design of their laboratories to accommodate the latest discoveries. The French context offers the most representative cases of chemical laboratories before the discovery of gases.

One of most famous successful teachers in chemistry of the eighteenth century was the Parisian apothecary Guillaume-François Rouelle (Partington 1962: 73–6). He delivered courses in the amphitheatre of the Jardin du Roi where he was employed from 1742 as a demonstrator of chemistry. His laboratory was some distance from the amphitheatre where he delivered his lectures, and for this reason the experimental demonstrations he was able to deliver on stage were necessarily limited. Nevertheless, his courses became so popular that they were attended every year by hundreds of participants – including foreigners and women – and as a result he gained unprecedented prestige.

However, it was in his shops in the Place Maubert – and after 1746 in the Rue Jacob – that Roulle delivered private advanced courses for selected students, in which more complex experiments were carried out. Hence Rouelle divided his professional life between two establishments: at the Jardin du Roi he lectured before large audiences (sometimes drawing several hundred students at a time), introducing them to the notions of vegetable, animal, and mineral chemistry, conducting chemical experiments, and even demonstrating a few scientific tricks for their diversion, while at his private laboratory he met a handful of students who wished to pursue chemistry at a higher level. This private and highly selective course focused on systematic experimental research, without the spectacle that he inserted to enliven his lectures at the Jardin du Roi. On Sundays Rouelle delivered lectures in his laboratory to an illustrious group of friends and colleagues, among whom Réaumur, Bernard de Jussieu, Henri-Louis Duhamel du Monceau, and François Venel were regular participants. During the early 1760s the young Antoine-Laurent Lavoisier also attended the private courses that were given in the Rue Jacob.

We have a detailed description of Rouelle's private laboratory made in 1747 by the Swedish mineralogist Sven Rinman:

> The whole building is approximately 50 feet (15 m) long and 18 feet (5.5 m) wide, consists of three rooms, the middle one being the biggest, 30 feet (9 m) long and with a stove against one side, 16 feet (4.7 m) wide and 4 feet (1.2 m) deep, so that the hood is somewhat above it.
>
> (Beretta 2011: 367)

Rinman wrote that the apparatus consisted of a forge hearth with its own little stove wall and a chimney; a small central reverberatory furnace; a larger air furnace for "big melts," in which the heat becomes quite strong by means of a dome mounted on top; distilling and sublimation furnaces; a large smelting furnace for copper and lead ores; and another small glass furnace for artisanal production of ceramics and enameling, as well as several portable furnaces, sand baths, and two distilling boilers. Rouelle's laboratory was relatively small and most of the apparatus seems to have been traditional.

Partly due to the huge success of Rouelle's chemical courses, during the 1750s and early 1760s there was a considerable development of laboratories, apparatus, and experiments. This progress can clearly be seen in the eighteen plates of the chemistry section in the twentieth volume of the *Encyclopédie* published in 1763 by Denis Diderot and Jean d'Alembert (Lehman and Pepin 2009). The first plate (Figure 3.1) depicts a chemical laboratory; it is likely that Diderot was illustrating the most famous chemical laboratory in Paris, namely Rouelle's, as he had been a pupil of the master. The scene shows all the main elements of an eighteenth-century laboratory; the most innovative feature is the introduction of a large chemical table upon which were performed all the chemical experiments and reactions that could be done without the help of violent fire (Morris 2015: 52–3). On it there are glasses and funnels used to filter liquors, glass jars with different kinds of stoppers, a small portable furnace, and other receivers.

At the left of the table a chemist is preparing a metallic solution while conversing with a *physicien* (physicist). The chemist is the one manipulating matter while the physicist is just conversing with him; this juxtaposition of roles underlines the superiority of chemistry as a science that, thanks to experimentation, goes beyond the surface of matter and reveals its inner

FIGURE 3.1 A chemical laboratory, from an engraving in the *Encyclopédie*, 1763. Courtesy the ARTFL Encyclopédie Project, University of Chicago.

structure. In addition to the chemist and the physicist there are three laboratory assistants and another chemist: the chemist in the background on the right is controlling the vapours coming from the detonation of nitre; the laboratory assistant on the left is carrying fuel from the charcoal storeroom on a lower floor to feed the furnaces; another assistant in front of the table is collecting water from a basin, and one on the right is washing a vessel in a barrel filled with water.

The simple operations performed by the assistants underline the crucial importance of water and fuel. Equally important is the ventilation of the room, as shown by the large chimney and the bellows. From left to right there is an *athanor* (a furnace that keeps the heat constant), an assaying furnace, and another furnace connected to a large retort. At the top of the mantelpiece we find a variety of glass receivers, retorts, alembics, and alchemical vessels used for slow distillation, such as the pelican, and, on the extreme right, an apparatus for measuring the quantity of air escaping from fermenting bodies. In addition to the description of the laboratory, Diderot added several engravings illustrating other types of chemical apparatus, most of which were drawn from Rouelle's own equipment.

In all, Diderot listed 265 pieces of apparatus. Among the most important instruments he included were various furnaces for smelting, distillation, decoction, calcination, and glassworking, as well as reverberatory furnaces, ovens of various type, stoves, equipment for *bains marie*, the apparatus for ascertaining the quantity of air which emanates from a body, an assay balance made by Gallonde, as well as all sorts of alembics, flasks, retorts, balloons, glass receivers, mortars, and various smaller utensils. While the origin of the depicted apparatus was mostly Rouelle's laboratory, the original designs on which they were based were often the work of other chemists, physicists, and alchemists. The pneumatic apparatus, for instance, was a modified version of that described by Stephen Hales in his *Vegetable Staticks* (1727); several furnaces were inspired by Johan Rudolf Glauber's *Furni novi philosophici* (1646–47), the comprehensive work on the subject. A considerable number of the instruments were reminiscent of the alchemical tradition, a revealing sign that eighteenth-century chemistry was a body of knowledge that continued to profit from the contributions of older traditions coming from outside academia or the artisanal trades.

The clarity and structure of the plates published by Diderot in the *Encyclopédie* showed that a detailed description of the apparatus and the laboratory could provide chemists with useful instructions. However, laboratory life was not easy, and experimentation was full of unexpected and often unpleasant disappointments. Moreover, the cost of maintaining a laboratory in good order was considerable, glassware often broke, clay was rapidly consumed, chemicals were not always readily at hand, and some of them, such as mercury, were extremely expensive.

In 1766 Pierre-Joseph Macquer (Partington 1962: 80–90), a physician who studied chemistry with Rouelle, published a *Dictionnaire de chymie*, which was translated into English, German, and Italian, and became the standard reference in European chemistry for the next decade. Macquer was a highly experienced and versatile chemist. He worked for the Bureau de Commerce to solve the technical problems relating to dyestuffs and, since 1757, he had been a consultant at the famous porcelain factory at Sèvres, for which he developed a porcelain furnace which was soon adopted as a standard throughout France.

Since the mid-1750s Macquer had also worked in the laboratory of the apothecary Antoine Baumé in the Rue Saint-Denis where they delivered a course in chemistry for several years. Baumé, who transferred his laboratory to the Rue Coquillière in 1762, was one of the most important suppliers of chemicals and chemical apparatus in Paris, and in addition to his shop he had several specialized workshops and laboratories (Davy 1955: 19–26). With his extended experience with chemical manufacturers and artisanal, teaching, and research laboratories, Macquer was a distinguished academic chemist with a long record of publications, in both practical and theoretical chemistry. Despite its lack of illustrations, the entry "laboratory" in Macquer's *Dictionnaire de chymie* was in fact far more informative than the one published by Diderot.

According to Macquer, chemistry was a science entirely based on experience, so it was impossible to understand it properly without a thorough knowledge of the instruments and apparatus needed to perform it. Macquer's effort to describe the ideal laboratory "proper for a philosophical chemist" and hence to avoid references to shops of the apothecary, the workshops of the artisans, or the large manufactories, was an important advance in comparison to the entry of the *Encyclopédie*. Although many characteristics and apparatus were common to all these sites, Macquer believed that the chemical laboratory had finally become a scientific site in its own right.

He recommended that the laboratory be constructed above ground level. Although a ground-level situation was "most convenient, for the sake of water, pounding, washing, etc.," the inconvenience caused by moisture often compromised the outcome of long experiments. Moreover, "In such a place, most saline matters become moist in time; the labels fall off, or are effaced; the bellows rot; the metals rust; the furnaces moulder, and every thing almost spoils. A laboratory therefore is more advantageously placed above than below the ground, that it may be as dry as is possible" (Macquer 1777).

The ventilation of the laboratory was also very important because of the exhalations and fumes, which could be extremely toxic. Macquer was among the first chemists to list, throughout the entries of his *Dictionnaire*, the dangers of chemical experimentation, and to suggest the necessary precautions to be taken in order to avoid accidents; among them, a well-aired laboratory was certainly the first and most important. But even less harmful fumes and dust such as that caused by charcoal had to be carefully handled so the laboratory could be kept tidy. Macquer also insisted on the importance of shelves for the orderly storage of glassware and chemicals. "In a laboratory where many experiments [are] made, one [could] not have too many shelves" (Macquer 1777).

The chemical table introduced by Rouelle soon became an essential feature of the chemical laboratory (Morris 2015: 52–3). Macquer placed it in the middle of the room and declared that it served for all the "preparations for operations, solutions, precipitations, small filtrations; in a word, whatever [did] not require fire, excepting that of a lamp." Macquer, like Rouelle, placed most of the furnaces under the large fireplace. His laboratory, however, contained only some of the furnaces listed in the *Encyclopédie*; the ones that could be traced to the alchemical tradition were left out. The glassware and most of the instruments were identical to those listed by Diderot.

Although glassware was of the greatest importance in chemical experimentation, Parisian chemists were unable to find vessels with the necessary resistance both to heat and cold. Macquer remarked:

> Vessels intended for chemical operations should, to be perfect, be able to bear without breaking the sudden application of great heat and cold, be impenetrable to every substance and inalterable to any solvent, be unvitrifiable and capable of enduring the most violent fire without fusing. But up to the present no vessels are known which combine all of these qualities.
>
> (Macquer 1753: 297)

By 1765 French academic chemists and scientists had some interest in glassmaking, but did not develop a consistent theory of its composition; in fact, glass itself was not yet a proper subject of chemical inquiry.

In addition to the apparatus and the utensils Macquer also included a long list of chemicals: metals, semimetals, vitriolic and nitric acids, common salt, vinegar, limewater, quicklime, various alkali, "refined essential oil of turpentine; oil of olives; soap, galls, syrup of violets; tincture of turnesol [litmus], or turnesol in rags; fine blue paper; river or distilled rain-water" (Macquer 1777). He also recommended the preparation of neutral salts, which were most commonly used during chemical operations.

The presence of so much equipment and chemicals led Macquer to insist on a cautious approach to experimentation, and to be methodical. Chemists had to be:

> well persuaded that method, order, and cleanliness, are essentially necessary in a chemical laboratory. Every vessel and utensil ought to be well cleaned as often as it is used, and put again into its place: labels ought to be fastened upon all the substances. These cares, which seem to be trifling, are however very

fatiguing and tedious; but they also are very important, though frequently little observed. When a person is keenly engaged, experiments succeed each other quickly; some seem nearly to decide the matter, and others suggest new ideas: he cannot but proceed to them immediately, and he is led from one to another: he thinks he may easily know again the products of the first experiments, and therefore he does not take time to put them in order: he prosecutes with eagerness the experiments which he has last thought of; and, in the mean time, the vessels employed, the glasses and bottles filled, so accumulate, that he cannot any longer distinguish them; or, at least, he is uncertain concerning many of his former products. This evil is increased if a new series of operations succeed, and occupy all the laboratory; or if he is obliged to quit it for some time: everything then goes into confusion. Thence it frequently happens that he loses the fruits of much labor, and that he must throw away almost all the products of his experiments.

(Macquer 1777)

Macquer's detailed description of the chaotic output of the complex and uncertain experiments carried out in the chemical laboratory reveal that the degree of accuracy of the apparatus was still rather far from the standards he to which he aspired.

The time for a radical reform of both the apparatus and the way to conduct experiments was coming. Before illustrating the discoveries leading to the transformation of chemical laboratories, it is necessary to briefly survey the sites created by the Parisian apothecary Antoine Baumé, whose early career, as mentioned, was associated with that of Macquer. Baumé represents a key figure of transition between the traditional chemical laboratory and the advent of a new phase of chemical experimentation.

Like Rouelle, Baumé was an ambitious apothecary who was confident in the role of both scientist and practitioner. He overcame the inferior social status conferred to pharmacists by associating himself with Macquer, who was a distinguished member of the Académie des Sciences, as well as of other important university and state institutions. In 1758 Baumé and Macquer signed a contract to deliver an advanced course in the laboratory that Baumé had opened in the Rue Saint-Denis in 1752. The contract covered the acquisition of the necessary equipment, the shared authorship of the discoveries made in the laboratory, and the project of joint publications. Although, like Rouelle at the Jardin du Roi, Baumé acted as the demonstrator, his role was nearly at the same level as that played by Macquer.

The courses were delivered regularly until 1772, when Baumé succeeded in being elected as a member of the chemistry section of the Académie des Sciences. As the number of experiments rapidly grew, Baumé moved his laboratory into a far more spacious building in the Rue Coquillère in 1762, where he was soon able to build five different laboratories, one of which was devoted to the preparation of drugs and another to the large-scale production of several products: lead acetate, compounds of antimony, mercurial salts, muriate of tin (stannous chloride), sal ammoniac (ammonium chloride), Glauber's salt (sodium sulfate), and volatile oils (Davy 1955: 19–21). Baumé was also a supplier of hundreds of medical remedies, some consisting of ingredients imported from the French colonies in the Americas, and despite the municipal prohibition to advertise the sale of commodities (Bourvet 1937: 236), in 1775 he published a comprehensive catalog of the drugs he kept in his shop, with their prices (Baumé 1775).

In addition to providing his clientele with remedies and chemicals, Baumé was also a major supplier of chemical instruments and apparatus. As early as 1768 he advertised in the *Avantcoureur* the sale of a chemical hydrometer of his own design, an instrument which remained in use up to the twentieth century. He sold all kinds of utensils, glassware, retorts, alembics, cucurbits, and furnaces. Baumé's trade – closely connected with the activity of his own laboratory – is quite important because he helped to improve and standardize the equipment of Parisian and provincial chemical laboratories.

The difficulty in carrying out this process of standardization was due to the variety of artisans and guilds involved in the constructions of apparatus. In 1773 Baumé gave a detailed list of the Parisian makers involved (Baumé 1773, vol. 1: cxxviii–cxlvi). Most of the glassware (alembics, retorts, phials, receivers, etc.) could be bought from pottery makers (*faïencers*) or from glaziers (*émailleurs*); earthenware from potters (*marchands potiers de terre*); furnaces and crucibles from the furnace makers (*fournalistes*) or from the foundrymen who were in charge of the production of the pipes. As the German crucibles were already regarded as the most resistant ones, there were tradesman importing them into France (Martinón-Torres 2006) and it was only in the mid-1780s that these crucibles began to be substituted by the more heat-resistant ones developed by Josiah Wedgwood in England. Instruments like copper receivers, alembics, and serpentines were made by boilermakers (*chaudronniers*); in 1777 Baumé himself made a 3.3-meter-high copper distilling apparatus, the largest then in Paris. Many useful apparatus and common utensils could be found at the ironmongers (*quincailleries*) and tinsmiths made different kinds of receivers. All balances – including assay balances and precision balances – were made by the balance-maker (*balancier*). The only "chemical" devices made in the workshops of professional instrument makers were thermometers, hydrometers, hygrometers, and pyrometers. Simpler apparatus such as marble and porphyry mortars, spatulas, scissors, pliers, wooden presses, and filters could be easily found in different shops.

The artisans involved in the building of a professional laboratory belonged to professional guilds, and so it was not easy to find the necessary expertise to satisfy the needs of chemical experimentation. It is significant in this regard that some chemists began to learn glassblowing to produce chemical glassware of different size and thickness. The dependence of chemists and pharmacists on the skills of artisans who knew little about science was a great obstacle to progress in chemical experimentation. For the chemist the laboratory was the main site of knowledge production, and without the decisive improvement of its apparatus and the management of its organization, it would have been extremely difficult to make any significant breakthroughs.

With the discovery of gases, chemists were forced to develop a different approach both to their traditional conception of apparatus, and to the philosophy of matter that had hitherto underpinned their chemical experiments.

GAS ENTERS THE LABORATORY

The discovery of the active role of gases in chemical reaction radically transformed apparatus and analytical methods. It is therefore significant in this regard that during the late 1770s and 1780s crucibles, glassware, and the most sophisticated equipment began to be made by professional instrument makers, or by entrepreneurs such as Josiah Wedgwood who had first-hand knowledge of chemical processes.

In a series of quantitative experiments made in 1753 and 1755 the chemist Joseph Black discovered what he called fixed air (carbon dioxide; Donovan 1975; Anderson 2015). The chemical laboratory of the University of Edinburgh was far from impressive, but Black was immediately concerned to enhance his discovery by adopting a high degree of precision in chemical analysis and to improve the quality of the equipment. Although little of his apparatus has survived (Anderson 1978: 20–33), he was among the first chemists to use the precision balance systematically and to seek able instrument makers – among them James Watt – to enrich his laboratory with accurate and innovative instruments.

Following Boerhaave's pioneering use of the thermometer, Black brought the effort to quantify the phenomena related to heat to the highest degree of perfection, and it was on the basis of the results obtained through his experiments that Watt's steam machine was built. Black's successor, Thomas Charles Hope, declared that: "Black's left a very excellent apparatus, and a considerable collection of mineralogy" (Anderson 1978: 58). Black's striving for experimental accuracy is not surprising; as early as 1753 he viewed chemistry as "a branch of natural philosophy" (Black 2012, vol. 1: 109).

Although the impact of Black's pneumatic discovery was not immediate, by the mid-1760s several scientists and artisans were inspired by his investigations. Significantly, the earliest protagonists of pneumatic chemistry were neither physicians nor pharmacists; most of them were natural philosophers. The analysis, identification, and classification of gases could be carried out only by those who were familiar with accurate apparatus and standard of precision, which by the early 1760s could be found in the "physical cabinet" rather than in the chemical laboratory. During the late 1760s and early 1770s the most important experiments and discoveries in pneumatic chemistry were made by Henry Cavendish and Joseph Priestley, neither of whom was a professional chemist. Cavendish was a wealthy aristocrat and distinguished himself by his extensive research in experimental physics, mathematics, and astronomy (Jungnickel and McCormmach 1999). He was also a skillful inventor of scientific instruments, constantly seeking to improve the standard of accuracy used during experiments. One of his first publications (1766) was devoted to the analysis of "factitious air," in which he presented, among other things, his discovery of an inflammable air (hydrogen). The Royal Society immediately awarded Cavendish the Copley Medal for this discovery. Cavendish continued to follow the progress made in pneumatic chemistry until the late 1780s. He contributed to this new discipline with several important discoveries, such as the experiments leading to the identification of the compound nature of water (Miller 2004), as well as with the decisive improvement of existing apparatus such as the eudiometer (a device invented by Marsilio Landriani in 1775 aimed at measuring the quantity of oxygen in closed vessels). Despite these outstanding contributions, Cavendish never regarded himself as a chemist, and his solitary habit of performing research and experiments prevented him from taking part in the controversies on the nature of elastic fluids that developed during the late 1770s and early 1780s. However, both Cavendish's laboratory and apparatus were known and admired by his contemporaries. His favorite site of research was a suburban villa at Clapham, now part of London, which was occupied by workshops, a laboratory, a forge, and an astronomical observatory (Wilson 1851: 164). The superior quality and precision of Cavendish's apparatus survived him, and a significant part of his equipment was still being used by Humphry Davy well into the early nineteenth century.

Joseph Priestley's background also had very little to do with chemistry, and his interest in the nature of gases was the effect of his extensive experimental work in experimental physics and electricity (Schofield 2004). In 1772 he presented the Royal Society with his *Observations on Different Kinds of Air*, which earned him the Copley Medal. In the years that followed, Priestley concentrated systematically on this topic, performed a considerable number of experiments, and improved existing apparatus. As he candidly admitted, he was not a "practical chemist," and he had little experience with chemical laboratories (Priestley 1775: 26), but in 1774 and 1775 he published further *Experiments and Observations on Different Kinds of Air*, and devoted much attention in describing the details of his experiments and of the apparatus he used.

Priestley's apparatus were not as simple as he claimed in the preface of his book. In order to manipulate gases he perfected Hales' earthenware pneumatic trough, later made of wood, which was filled with water and used to collect and manipulate different gases with the help of cylindrical glass jars. The shape of glass jars, receivers, tubes, and glass balloons was adapted to the need to store gases, and for this purpose special attention was paid to the construction of joints, stoppers, and taps. Significantly, all this glassware was not supplied by an ordinary glassmaker, but by William Parker, a renowned instrument maker of London.

Priestley illustrated other apparatus in his works: a more sophisticated device for expelling gases from solids; an apparatus for impregnating a fluid with gas; a eudiometer; and an apparatus for making electric sparks in any kind of gas. Priestley's background in natural philosophy inspired him to use a powerful burning lens made by William Parker to heat *mercurius calcinatus per se* (mercuric oxide), which led to the discovery of what Priestley called dephlogisticated air (oxygen). The use of burning lenses of various sizes became quite popular during the 1770s, and several experiments were made to submit precious stones, especially diamonds, to the action of intense heat (Lehman 2016). Priestley also adapted other physical instruments, such as electrical machines, barometers, thermometers, hygrometers, air pumps, eudiometers, and microscopes for pneumatic investigations; because these instruments were more sensitive than ordinary apparatus they were kept in a separate room. Towards the end of his life Priestley had a laboratory consisting of several rooms equipped with elaborate and expensive instruments. As he was used to conducting experiments with no preconceived plan, Priestley was not always able to interpret his own results easily. The difficulty he had classifying gases within natural philosophy made it impossible for Priestley to interpret their role in chemical reactions.

The first professional chemist to fully adapt his laboratory to the development of pneumatic chemistry was the chemist Torbern Bergman (Partington 1962: 179–99; Bergman 1965). When Bergman was appointed professor of chemistry at Uppsala University in 1767, he did not appear to have any relevant qualifications for the post. In his early years as a student, he showed such passion for mathematics and astronomy that a short incursion in natural history under the supervision of Carl Linnaeus, the celebrity of the university, did not persuade him to change his original plan of study. Thus, in 1758, he defended a thesis entitled *De interpolatione astronomica*, which was soon followed by the Newton-inspired *De attractione universali*. These scientific treatises earned him the position of associate professor of mathematics in 1760, and during the early 1760s Bergman took a keen interest in meteorology, electricity, and in the electrical properties of tourmaline. As pointed out by his pupil Peter Jacob Hjelm, Bergman's background as "a mathematician, astronomer, naturalist and physicist" was not in conflict with his appointment as a professor of chemistry (1786: 39). On the contrary, the time was ripe to overcome the traditional disciplinary boundaries of chemistry. Bergman soon understood that pneumatic chemistry represented the most important branch of the science, and blamed his contemporaries for not seeing its relevance. As early as 1774 he determined the specific gravity of carbonic acid gas (carbon dioxide), which he named aerial acid. His method of analysis impressed Priestley and many other European scientists involved in pneumatic researches. In 1778 he extended his quantitative method to the analysis of mineral waters.

Among Begman's relevant early innovations, Hjelm has discussed the redesign of the chemical laboratory following a fire. As it took nearly a year and a half to have the building rebuilt, Bergman gave a glimpse of the ways he wished to transform chemistry teaching and research during the first course he held in the *Auditorium oeconomicum*, where he had displayed Anton von Swab's collection of minerals (Hjelm 1786: 41). Mineral collections were relatively common throughout Europe (Wilson 1994), but Bergman was among the first to use them systematically in a chemical laboratory. Although following the Linnaean tradition minerals were studied within the discipline of natural history and classified according to their external characters, Bergman arranged the specimens of his collection "according to their content and composition," thus putting great emphasis on the role of chemical analysis.

The chemical constituents of minerals had already been explored in the *Laboratorium chemicum* of the Board of Mines. As early as 1758 Axel Fredrik Cronstedt, a distinguished member of the Royal Swedish Academy of Sciences and a teacher at the Board of Mines (*Bergskollegium*) in Stockholm, had published a textbook that met with great success despite being written in Swedish. In the *Försök till mineralogie* (1758), Cronstedt specifically brought mineralogy outside the sphere of natural history; he abandoned the hitherto dominant traditional descriptive and taxonomical system of classifying minerals, preferring to rely on the use of the blowpipe for chemical analysis (Jensen 1986). Although the blowpipe had been used outside Sweden also, thanks to Cronstedt and Bergman the chemical analysis of minerals became a priority during the early 1760s and 1770s. Bergman's analytical method was the object of his successful chemical courses, and it rapidly spread among his most talented students. One of them, the Spaniard Juan José de Elhuyar, exported it to the mines of South America, where in 1784 he was pleased to notify his mentor of his discovery of a new metal (tungsten).

Bergman appropriated the blowpipe method used by the Board of Mines and adapted it to academic chemistry (Bergman 1779; Bergman 1782). This meant that instructions for the use of the blowpipe were made clear to all potential users, and that the focus of the experiments was not driven exclusively by practical and economic needs but by an ambitious design to reform methods of traditional dry chemical analysis. So Swab's mineral collection was not a mere ornament of Bergman's laboratory but an asset which allowed him and his students to gain authority on a new experimental field, and to compete with the *Laboratorium chemicum* in Stockholm.

Besides the gallery with minerals, Bergman had a room filled with very precise didactic models of "the apparatus which is used for manipulation of metals" and of furnaces and instruments related to chemical arts such as "the manufacture of glass, porcelain, glazed earthenware, pottery vessels, brick, tobacco pipes, all kind of salts, oils, gunpowder, lampblack [and] all fire-resistant dyes" (Bergman 1985: 470). Bergman took inspiration in using models of chemical and mining apparatus from his colleague, the professor of economics Anders Berch, chair of the building in which Bergman took his first course in chemistry. The utilitarian philosophy embraced by Berch was enhanced during the courses he delivered in political economy by the extensive use of models of machines illustrating the technical advancement in manufacturing; this approach was in accord with Bergman's need to introduce his students to chemical arts from a general perspective.

In addition to the minerals, the models and a "cupboard of noteworthy chemicals" such as salts, metals, and acids, Bergman's laboratory consisted of a forge with a good double bellows, reverberatory and glassworking ovens, an *athanor* (constant-temperature oven) for digestion, open and closed boilers for distillations, an oven for coloring and salt boiling in which a copper cauldron was bricked in as a container, and two cast-iron cauldrons working as sand baths in which two loose cylindrical pots, one made of fine tinplate, the other of lead, could be put when needed for colors and various evaporations. There were several contrivances on which to hang distillation vessels over an open fire in small, open ovens, meaning, as Bergman remarked, one could "conveniently see all that [took] place in a vessel from beginning to end" (Bergman 1985: 471).

Bergman also designed different kinds of blowpipes and owned a remarkably rich collection of vessels and receivers (made of glass, refractory clay, and metals), mortars, and other more common equipment. As Priestley had done, Bergman arranged a special room in his laboratory for physical instruments such as hydrometers, an electrical machine, two electrophores (electrostatic generators), a hygrometer, an air pump, various balances, a compound microscope, various kinds of barometers and thermometers, and special glassware. A comprehensive list of both fine instruments and chemical apparatus was compiled soon after Bergman's death by his successor in the chair of chemistry, Johan Afzelius.[2] Bergman justified the quantitative and qualitative expansion of laboratory equipment by the growing complexity of experimentation and the need to maximize its accuracy in order to avoid or overcome useless disputes. His laboratory embodied in one site both innovative experimental research and the didactic side of chemistry. The combination of research and didactics proved to be extremely successful, and it was emulated by several chemists throughout Europe.

Bergman's new design of the university chemical laboratory has been often contrasted with the extraordinary experimental skill displayed by Carl Wilhelm Scheele during the mid-1770s in his small apothecary shop in Köping, Sweden. Scheele is commonly regarded as the most prolific chemist of the eighteenth century, and his remarkable number of discoveries, that of oxygen above all, was the object of universal admiration by his contemporaries (Partington 1962: 204–34). The physical site of Scheele's discoveries, however, has never been the object of a systematic study. During the late 1760s he worked in an apothecary shop in Stockholm where he had little time to experiment. This situation changed dramatically when he moved to Uppsala in 1770. There he was welcomed into Bergman's laboratory, where, together with Johan Gottlob Gahn, he made many experiments in different research fields. Scheele benefited from the fine equipment he had at his disposal, while Bergman profited from the results obtained by Scheele in his laboratory. As Bergman kept his large and up-to-date library in the laboratory, it was no doubt here that Scheele was made aware of the latest progress made in pneumatic chemistry in Europe and, for instance, learned about the recent discoveries made in France.

Indeed, Scheele's famous letter, addressed in September 1774, to Lavoisier containing his method for producing oxygen was translated into French and written by Bergman himself (Boklund 1957). It was Bergman who also eventually found a suitable publisher for Scheele's treatise *Chemische Abhandlung von der Luft und dem Feuer* (Uppsala 1777) and wrote a preface to it endorsing Scheele's chemical authority. It is not surprising that Scheele's career in chemistry took off in Bergman's laboratory, because nowhere else could he have had such a rich collection of chemicals, minerals, equipment, and instruments at his disposal – which he missed and regretted when he moved to Köping in 1775.

Scheele's and Gahn's attendance in Bergman's laboratory is interesting because neither of them were academics, but after their acquaintance with Bergman they began to publish academic papers and gained a reputation within the national and European chemical community. Although Scheele and Gahn were both exceptionally capable for different reasons, their debt towards Bergman's laboratory is clear. Moreover, they were far from alone; many of Bergman's numerous students were in fact engaged in laboratory research, and a few of them, such as Johan Afzelius, Johan Gadolin, and Peter Jacob Hjelm, later distinguished themselves as skillful chemists, thus creating the basis for a reputable Swedish school of chemical research.

Bergman's laboratory was attended by hundreds of students, many from foreign countries, and it soon became a model followed outside Sweden. It was in Paris that the transition between the old chemical workshop and a highly sophisticated laboratory was consummated (Beretta 2009; Beretta and Brenni forthcoming). Antoine-Laurent Lavoisier, who is usually credited for the construction of a unique and extravagant chemical laboratory (Poirier 1996), in fact followed the path inaugurated by Bergman, a chemist he greatly admired. Lavoisier, like Bergman, had preferred experimental physics in his youth, and served his chemical apprenticeship during mineralogical travels made between 1763 and 1767 with his mentor Jean-Etienne Guettard. It was during these trips that he conceived of portable instruments to perform experimental work outside the wall of the chemical laboratory. For example, Lavoisier invented a new type of portable aerometer with which he made the analysis of mineral waters, and also commissioned other portable instruments such as barometers and thermometers, which he used during his mineralogical travels. Although Lavoisier continued to perform open-air experiments for the rest of his life, his fame is justly connected with the creation of the chemical laboratory at the Arsenal in Paris.

Lavoisier took up residence in the Petit Arsenal in spring 1776. It was in this spacious building that Lavoisier gradually put together a laboratory with more than 10,000 pieces of apparatus. The area consisted of the l'Hôtel de la Régie des Poudres & Salpêtre in the Rue des Ormes, a building facing the Cour du Salpêtre, probably occupied by Lavoisier and the work-rooms for the production and refining of saltpeter (Belhoste 2011: 207–17). In addition to this there was a public garden with nearby access to water at the Fossées de l'Arsenal, where a gunpowder warehouse was located. The laboratory consisted of a building of two floors, a courtyard, warehouses, a garden, and a source of water. On the ground floor the laboratory consisted of three rooms. In the center was the room for chemical experiments, which communicated with two rooms, one with easy access to large quantities of water, the second used as a storeroom for the gasometer, the eudiometers, the electrical machines, the pneumatic pumps, the burning mirrors, the barometers, the thermometers, the balances, the calorimeter, the hydrometers, and more generally for all the apparatus which could be attacked by acid fumes and thus damaged. A room on the upper floor held all the chemicals that were kept well in good order, as Macquer had prescribed. Although there are no contemporary records, it is likely that Lavoisier, like Bergman, kept his large collection of minerals (about 3,000 items) in a separate room.

Lavoisier used every available space in the Arsenal to make experiments. In June 1781 in the Arsenal's garden he conducted experiments with Laplace on the thermal expansion of materials, using a new optical dilatometer. In March of 1782 Alessandro Volta, Laplace, and Lavoisier used the garden once more to carry out experiments on the vaporization of fluids, using Volta's condenser. The garden also became the site of experiments commissioned by the Académie des Sciences on new weights and measures. Lavoisier also undertook chemical analysis of the water of the nearby channel, hoping to collect inflammable gases.

Lavoisier's laboratory underwent several architectural changes. The Arsenal could also make use of a small nearby workshop. Lavoisier's most promising assistant, Armand Séguin, lived close to the Arsenal, in Faubourg St. Antoine, and he had a workshop in his home where together with Lavoisier he carried

LABORATORIES AND TECHNOLOGY

out experiments, such as those on the fusion of platinum in July 1791. The English agronomist Arthur Young was impressed by Lavoisier's collection of physical instruments, and in particular the gasometer, the apparatus used in experiments on the composition and decomposition of water:

> That apartment, the operations of which have been rendered so interesting to the philosophical world, I had pleasure in viewing. In the apparatus for aerial experiments, nothing makes so great a figure as the machine for burning inflammable and vital air, to make, or deposit water; it is a splendid machine … .
>
> (Young 1792: 64–5)

The first prototype of the gasometer was conceived by Lavoisier in 1782 (Figure 3.2). In collaboration with the engineer Jean Baptiste Meusnier de la Place, it was then improved by the Parisian instrument maker Pierre Bernard Mégnié, who took ninety-five days to perfect it. On February 28 and 29,

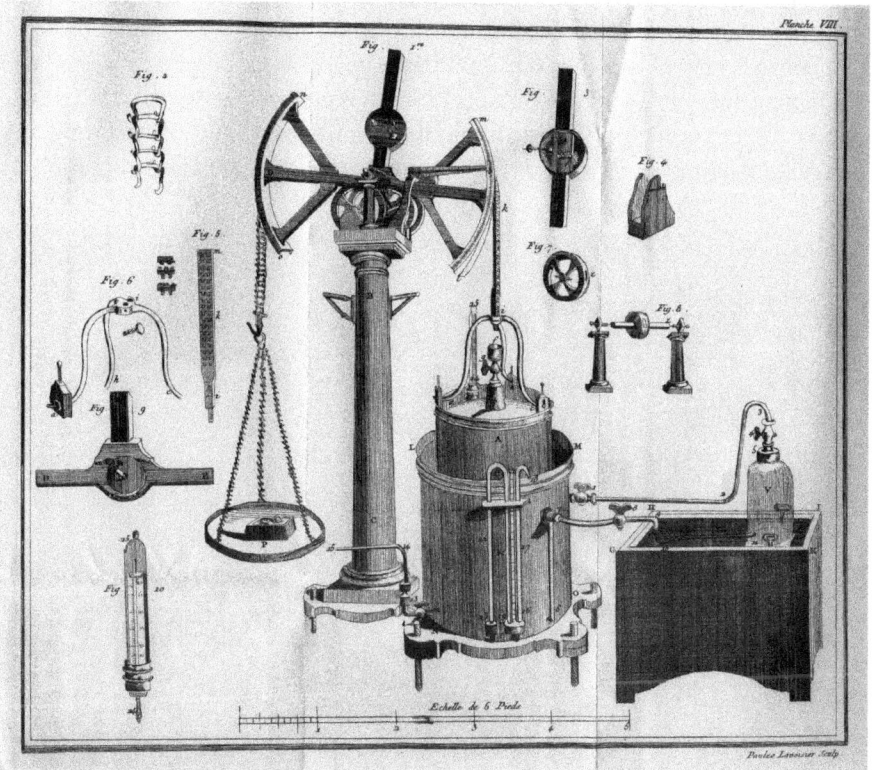

FIGURE 3.2 A.-L. Lavoisier's gasometer, from an engraving made by Marie Lavoisier and published in the second volume of A.-L. Lavoisier, *Traité élémentaire de chimie*, 1789. Author's copy.

1785 the instrument was used at the Arsenal during the spectacular public experiment on the analyses and synthesis of water. This large-scale experiment demonstrating that water was a compound of hydrogen and oxygen lasted several days and was witnessed by more than thirty savants, both foreign and Parisian. The experiment was extremely successful, and was repeated several times in the following months, supporting Lavoisier's theory of oxygen with unprecedented success. Despite the success, Lavoisier must have been dissatisfied with the apparatus, because just a few months later he ordered Mégnié to make two even more accurate gasometers. These gasometers, which cost 7,554 livres (Holmes and Levere 2000: 122), were the most expensive pieces of apparatus of his laboratory and by far the most expensive instrument ever conceived in the history of chemistry. Lavoisier justified its high cost by the need to bring into chemistry the same standards of precision adopted in exact science. The use of such a complex apparatus implied a radical transformation of the organization of laboratory life; Lavoisier involved a number of instrument makers, artisans, and assistants at the Arsenal, thus making chemical experimentatin a collective enterprise embodying different skills. Lavoisier looked for collaborations not only with ingenious young technicians, he also introduced a policy of credit recognition which favored the emergence of non-academic figures within academic contexts. This same organization of experimental works was repeated during the famous experiments on human respiration (1790– 1792), minutely described by Madame Lavoisier in two beautiful sepia drawings (Figure 3.3).

FIGURE 3.3 A.-L. Lavoisier in his laboratory conducting an experiment on the respiration of a resting man. Photogravure (ca.1850) after a drawing (ca. 1790) by Marie Lavoisier. Courtesy Wellcome Collection CC BY.

Lavoisier gave full credit to his laboratory assistant Armand Séguin, both for having conceived the apparatus, and for sharing his innovative interpretation of the results obtained with it.

CONCLUSION

Madame Lavoisier's drawings reveal a number of other interesting things that serve to illustrate how much chemical instrumentation and technologies had changed since the start of the eighteenth century. First and foremost, we owe it to Lavoisier that the traditional chemistry laboratory, previously not much larger than a kitchen, became so large that the experiments were carried out in different rooms. Second, in contrast to the public experiments performed by Rouelle and Fourcroy in amphitheaters, Lavoisier's sites of experimentation actively involved assistants, apprentices, and observers in a non-hierarchical way. Within this structure it became natural to encourage and help those working in the laboratory, who thus were able to reach original results and publish them in prestigious journals, such as the *Mémoires de l'Académie des Sciences* or the *Annales de chimie*. Last but not least, the apparatus displayed in Lavoisier's laboratory was no longer represented by a collection of standardized devices supplied by the Parisian guilds, but the result of the professionalization of skilful instrument makers, who, thanks to the growing demand for extremely sophisticated instruments, were actively involved in the process of chemical experimentation (Beretta 2014a). Lavoisier's laboratory operated as a true school for experimental apprenticeships, setting the model for the didactic laboratory, later successfully introduced at the École Polytechnique in 1795.

CHAPTER FOUR

Culture and Science: *Chemistry in its Golden Age*

BERNADETTE BENSAUDE-VINCENT

INTRODUCTION[1]

In a common depiction by historians of chemistry a generation or two ago, the eighteenth century was a rather dull period populated by obscure figures merely preparing the stage for Lavoisier's great achievements. In stark contrast to this Lavoisier-centered, backward-looking narrative that came to prevail among generations of chemists and historians of science, the historical figures themselves were inclined to consider their epoch the Golden Age of chemistry. Eighteenth-century chemists who self-consciously characterized their science in the introductions of their books drew a confident and approving image. For instance, Pierre Joseph Macquer described the context of publication of his successful *Dictionnaire de chimie* in these positive terms:

> We now have the advantage of seeing the best days of Chemistry. The taste of our age for philosophical matters, the protection of princes, the zeal of a multitude of illustrious and intelligent persons attached by inclination to the study of this science, the profound skill and ardour of modern Chemists, whom we do not attempt to praise, because they are above all eulogiums,

seem altogether to promise the most brilliant success. We have seen Chemistry drawing its origin from necessity, and receiving a flow and obscure increase from avarice. To true philosophy it was reserved to bring it to perfection.

(Macquer 1777: xviii)

Chemistry was depicted as a flourishing science attracting many adherents under the patronage of princes. It was regarded as an integral part of the Enlightenment culture that developed all over Europe.

To be sure, this self-proclaimed encomium cannot be taken at face value. Macquer's words appear as a typical rhetorical claim to promote a publication. However, there are a number of objective indicators testifying to the social and cultural prestige of chemistry in eighteenth-century Europe.[2] For one thing, chemistry enjoyed rising academic status throughout the eighteenth century. It became one of the six classes of the Royal Academy of Sciences of Paris when the Academy was reorganized in 1699. In the French provincial academies where enlightened amateurs cultivated "sciences, arts et belles lettres" (Roche 1988), chemistry was particularly important, especially in Montpellier, Bordeaux, Nancy, Dijon, and Rouen. All these academies encouraged research on chemical topics through competitions and awards.

Furthermore, chemistry was one of the disciplines taught in European universities. It became prominent in the curriculum of many universities from Uppsala and Edinburgh to Montpellier and Coimbra in the 1770s (see Chapter 7 in this volume). Although chemistry most often featured as an adjunct science to the medical curriculum, it was increasingly taught as an autonomous discipline, with a wide corpus of technical practices organized around a number of basic principles. In the Scottish universities, for instance, chemistry courses routinely reached an audience beyond the circle of medical students. Because William Cullen's lectures in Glasgow and Edinburgh treated chemical subjects in mineralogy, agriculture, textile production, and mineralogy, he attracted large audiences of gentlemen who contributed to the city's prosperity (Golinski 1999: 11–49). From 1766 to 1795, Cullen's successor, Joseph Black, had 3,765 registered students; by the 1790s the attendance in his classes often exceeded 300 (Christie 2015: 87–8). Although we do not know the precise figures, it seems likely that the audience for chemistry courses also increased in Swedish, Dutch, and German universities.

A PUBLIC TASTE FOR CHEMISTRY

According to Macquer, chemistry suited the tastes of the Enlightenment. Sciences in general played an important role in the Enlightenment movement, which highly valued reason as the primary source of legitimacy against the authority of religion and monarchy. Physics, chemistry, and geology were integral parts of what was known as "natural philosophy." This phrase was a synonym for the "experimental philosophy," emphasizing the departure from the medieval scholastic tradition of natural philosophy. Chemical experiments, which had been considered the province of mere "sooty empirics" in the seventeenth century (Clericuzio 2010), were highly praised by the middle of the eighteenth century.

To better understand what Macquer meant by the term "taste of the period," let us refer to the contemporary *Essay on Taste* (*Essai sur le goût dans les choses de la nature & de l'art*) by Montesquieu (Spector 2013). Montesquieu noted that our tastes proceed from pleasures of the body, from pleasures of the mind, as well as from social prejudices and habits. The taste for science, in general so pronounced in the mid-eighteenth century, combined intellectual curiosity with novel sensory experiences. It also involved civic or patriotic commitment to material improvements, an important value that was disseminated through diverse institutional, social, and cultural networks.

The mid-eighteenth century was a period when scientific knowledge and practices in general spread throughout culture and society (Stewart 1992; Golinski 1999). By the end of the century, Louis-Sébastien Mercier, a keen observer of Parisian culture, had the impression that "the reign of humanities is over, physicists replace poets and novelists, the electrical machine takes the place of a theatre play" (Mercier 1999). Natural phenomena such as electricity, magnetism, and novel airs provided sensational spectacles. A good indicator is the semantic shift of the canonical metaphor of nature as a theater. This image, introduced in 1686 by Louis-Bernard Le Bouyer de Fontenelle in a popular book, conveyed the urge to look metaphorically behind the stage at the opera machinery, at the hidden mechanisms of nature responsible for appearances in a Cartesian manner. The metaphor took an alternative meaning in the course of the eighteenth century, suggested by the bestseller *Le spectacle de la nature* published by Abbé Noel-Antoine Pluche in 1749. Pluche used the metaphor of the theater of nature to encourage his readers to enjoy the visual experience of natural phenomena. Just look at what is striking on the stage of nature, he urged, without asking to visit the machinery backstage (Pluche 1749: x). Science became a matter of spectacle (Bensaude-Vincent and Blondel 2008). Aristocratic salons, provincial academies, shops, and public fairs displayed instruments such as air pumps, electrical machines, colliding ivory balls, magic mirrors, and automata as sensory experiences and aesthetic objects.

While optics, mechanics, and electricity were attracting aristocratic or popular audiences as curiosities or entertainments, the public "taste for chemistry" was more widely spread because it also included trades. John Perkins's exploration (2010) of the audience of chemistry courses in various European towns show that a significant number of apothecaries, physicians, manufacturers, and gentlemen farmers devoted time and money to the culture of chemistry. While in the early decades of the century public lecturers on chemistry mainly recruited their audiences from medical doctors, surgeons, and apothecaries, they gradually attracted larger audiences with no vocational interests.

Significantly, Peter Shaw, who had been lecturing in London to the medical community, moved in 1733 to Scarborough, a resort town in Yorkshire renowned for its medicinal spa. His lectures covered practical topics such as pharmacy, winemaking, and mining, combined with spectacular demonstrations of the properties of phosphorus (Golinski 2003). In Paris, where chemistry had been taught at the Jardin du Roi since 1663, Guillaume-François Rouelle attracted large audiences in his public demonstrations of chemistry delivered from 1742 to 1763. Although the prospectus of his course promised wonders and entertainment, there was much more to be retained from his lectures (Roberts 2008). In fact, Rouelle delivered a full course dealing with the three realms of nature, and providing theoretical notions together with more or less predictable experiments. In addition to Rouelle's public lectures, Paris offered about thirty public and private courses for a fee in the mid-1780s. Three hundred people attended such courses in the 1730s; the audience reached 1,000 people in 1761, and 3,500 in 1786 (Perkins 2010).

Chemistry also attracted women interested in making perfumes, medicines, cosmetics, and cleaning products. Jean-Jacques Rousseau witnessed the passion of aristocratic women for chemistry during the time he spent with Madame de Warens. She performed many experiments at home, and was vulnerable to charlatans:

> Thus, although she had some principles of philosophy and physics, she did not fail to acquire the taste her father had for empirical medicine and for alchemy; she made elixirs, dyes, balms, medical precipitates, she claimed to have secrets. Taking advantage of her weakness, charlatans fell upon her, plagued her, ruined her and in the midst of furnaces and drugs used up her mind, her talents and her charms, with which she could have brought delight to the best societies.
>
> (Rousseau 1995: 42)

In the throngs who attended Rouelle's demonstrations, there were women seeking more reliable recipes and more robust knowledge who rubbed shoulders not only with medical doctors, artisans, manufacturers, and gentlemen farmers, but also with a new public of *philosophes*. The French writers known as the philosophes cultivated values such as rationality, liberty, tolerance, and public interest that came to epitomize the spirit of Enlightenment. Among them, Robert-Jacques Turgot, Denis Diderot, and Jean-Jacques Rousseau acquired chemical knowledge as part of their liberal education, and thus greatly contributed to the philosophical credentials of chemistry. Some of them set up private laboratories at home and performed experiments. Rousseau witnessed this aristocratic passion for chemistry again when he became the secretary of Louise Fontaine Dupin and her husband Claude Dupin, a wealthy tax collector, who had equipped a laboratory of chemistry in their castle at Chenonceaux. Rousseau was also in charge of the instruction of the couple's son Dupin de Francueil, and with his pupil he attended Rouelle's courses in 1742–1743.

In the 1770s, the discovery of new gases reinforced the public taste for chemistry. For instance, Joseph Priestley, who was deeply involved in pneumatic science, taught at two Dissenting Academies, Warrington and Hackney. In his public lectures he reproduced the experiments that had led him to the discovery of a number of gases with very simple apparatus and a detailed description of his procedures, thereby inspiring a whole generation of public lecturers to reproduce pneumatic experiments (Golinski 2003). He also performed experiments to more select audiences in the London home of his aristocratic patron, the reform-minded politician the Earl of Shelburne.

Many of those who attended such courses also purchased books. As print culture proliferated during the Enlightenment (Siskin and Warner 2010), a large number of treatises circulated that claimed to provide elementary notions of chemistry. Diderot and d'Alembert's *Encyclopédie* (1769) was not the only symbol of Enlightenment print culture (Darnton 1987). So pronounced was the taste for philosophical matters all over Europe that such large-scale editorial enterprises were very successful. Macquer's *Dictionnaire* proved a great success: both the 1766 edition and the 1778 enlarged edition were widely circulated and translated into many languages (Neville and Smeaton 1981). Interestingly, the translators of Macquer's *Dictionnaire* included notes and additions. They did not hesitate to discuss the contents of entries or to add new ones. They felt it was legitimate to contribute to the improvement of the original text. This hybrid activity of translating, editing, and augmenting the original text – characteristic of a period when there was no copyright (Johns 1998) – encouraged the proliferation of publications (some authorized, and some not) and the greater participation of enlightened amateurs in the advancement and dissemination of chemical knowledge.

How did chemistry gain such a high social profile in eighteenth-century Europe? Two driving forces of its cultural ascent will be outlined below: the rhetoric of a radical break from the past which shaped the image of chemistry as a serious and useful science, and the role of chemists in promoting a close alliance between concerns about nature and society, as expressed in the concept of "oeconomy." This chapter will then situate chemistry on the general map of knowledge by considering its changing relations with neighbor branches of natural philosophy, and the centrality of chemistry in a number of eighteenth-century philosophical works.

BREAKING WITH THE PAST

To enhance the status of their science, eighteenth-century academic chemists used the resource of historical discourse. By the middle of the century, most textbooks included a lengthy historical introduction aimed at situating chemistry in the *longue durée*, from the origins of chemistry to the emergence of modern science. The legacy of alchemy became a critical issue, because it both allowed and obliged chemists to reshape the cultural identity of their science.

While the two terms "alchemy" and "chemistry" were used interchangeably in the early eighteenth century, in the 1720s Fontenelle launched a campaign to characterize alchemy as a vain, fraudulent, and unscientific pursuit. He used his power as the perpetual secretary of the Paris Academy of Sciences to marginalize and even ban any attempts at performing metallic transmutations (chrysopoeia; Principe 2007). In particular, Fontenelle summoned Etienne-François Geoffroy, the successor of Wilhelm Homberg (who had sought to achieve metallic transmutations until his death in 1715) to officially condemn alchemical practices. The pamphlet entitled "*Des supercheries concernant la pierre philosophale*" (frauds concerning the philosophers' stone), published in 1722 under Geoffroy's signature, was in fact a paraphrase of an older text published in 1617 by Michael Maier to distinguish between true chrysopeians and charlatans.

Although the rhetoric of this dichotomy was stronger in Paris than in other places, most academic chemists adopted similar strategies; they did not hesitate to darken the picture of past practices in order to suggest that a radical change had taken place. In his inaugural lecture at the chair of chemistry in Leiden in 1718, Herman Boerhaave carefully marked his distance from the dirty activity of chemists tending their fires. Because his successful textbooks went through many pirate editions and translations, it is wise, as John Christie (1994) recommends, to distinguish between various versions. In the English translation by Peter Shaw entitled *New Method* (Shaw 1727), the historical narrative included speculations about the origin of chemistry, then a description of the corruption of the art due to the "enthusiastical speculations" of third-century alchemists, followed by a restoration initiated by Roger Bacon in the thirteenth century. Modern chemistry, he wrote, emerged from Paracelsus, who introduced the chemical reform of Galenic medicine, and from Jan Baptist van Helmont who developed the experimental method. Thus, in this early edition, Boerhaave claimed "a Paracelsian lineage for modern chemistry" and disregarded the changing theories of matter (Christie 1994). The French translation of his *Elements of Chemistry* significantly differed from this. It clearly stated that chemists had overcome their errors and learned to conduct careful experiments before making extended theoretical claims, thus turning chemistry into a definite science, recommendable for its utility (Boerhaave 1754: li–lxxj).

In the introduction to his *Dictionnaire de chimie*, Macquer depicted the close connection between experiment and reasoning as the distinctive feature of modern chemistry. Macquer went even further in his effort to reject the alchemical lineage of modern chemistry:

> Although we have advanced but little in the History of Chemistry, we cannot proceed without mentioning a singular madness which possessed Chemifts. It was a kind of epidemic malady, the symptoms of which mewed to what excess of folly the human mind could proceed, when influenced by a strong prepossession.
>
> (Macquer 1777: vol. 1, 373)

In Scotland, William Cullen's historical narrative also celebrated the alliance between experiment and theory in describing a renaissance of chemistry. However, Cullen clearly attributed the change to the introduction of mechanical and post-mechanical philosophy. Francis Bacon, Robert Boyle, and Isaac Newton were presented as the real founders of chemical philosophy (Christie 1994). Despite local dissimilarities, most historical narratives were increasingly oriented toward the claim of a radical break with the past that secured chemistry against all reproaches. This break also had a social aspect, as pointed out by John Robison. In assessing Cullen's reformulation of chemistry, he argued that Cullen's "philosophical chemistry," now aligned with the natural and moral philosophy of the arts curriculum at the University of Edinburgh, succeeded in "taking chemistry out of the hands of the artists [i.e. artisans], metallurgists and pharmacists, and exhibiting it as a liberal science, fit for gentleman to study" – a visible transformation of chemistry's social status (Robison 1803: xxii).

In view of such ubiquitous claims, Lavoisier's self-proclaimed revolution is striking precisely as a continuation of this rhetorical strategy. Just as earlier chemists advocated a "new chemistry," Lavoisier claimed that his research agenda had nothing to do with past chemistry. The self-proclaimed "founder" of a new chemistry was in fact re-enacting an older pattern. In so doing, he nevertheless pushed the rhetorical claims further, as he adopted a more radical stance (Bensaude-Vincent 1998). First, he introduced the phrase "revolution in physics and chemistry," which had not been used earlier.[3] Second, the term revolution referred to his own achievements rather than to a collective endeavor, as he claimed full credit for it. Finally, Lavoisier made a clean sweep of the past when in 1785 he urged his fellow chemists to forget about the phlogiston theory, and then four years later he banished history entirely from his *Traité élémentaire de chimie*.

BETWEEN NATURE AND SOCIETY

These rhetorical claims suggesting a radical break with the past and a fresh new start proved very powerful, because they were uncritically endorsed by generations of chemists who thus obscured a clearer view of chemistry in the eighteenth century. Yet even though they contributed to change the earlier popular image of chemistry as magic, necromancy, and poisons, and to secure in its place the image of a "true science," the eighteenth-century chemists' historical narratives were not by themselves sufficient to secure the social prestige of chemistry.

John Perkins (2013) persuasively drew historians' attention to the variety of sites where chemistry was pursued in eighteenth-century urban societies. If chemistry became central in eighteenth-century European culture, it is primarily because of the social and economic importance of trades related to chemistry such as glassmaking, metallurgy, mining, and smelting (Fors 2003), porcelain (Klein 2014a), dyes (Nieto-Galan 2001; Lehman 2012), tanning, pharmacy (Simon 2005), cosmetics (Lanoé 2003), hat-making (Guillerme 2007), and fertilizers. Because such activities promoted the public welfare, a number of artisans who attended chemistry courses and academic chemists interacting with them increasingly took on official positions.

> By forging social connections with practitioners and patrons of the arts and by the concrete work of experimental research, eighteenth-century chemists positioned their science at a pivot point in the relationship between natural knowledge and power over the material world, a relationship that was emerging as a characteristic of the age.
>
> (Golinski 2003: 380)

Indeed, chemists actively contributed to the emergence of the relationship between knowledge and power over the material world as a major feature of eighteenth-century culture. "Artisanal-scientific experts" (Klein 2012b) served as consultants, advisors, or inspectors for various state agencies such as the Swedish Bureau of Mines (Fors 2003), the mining administration of Prussia (Klein 2012b), and the French royal manufactures (Lehman 2012). Of special importance for the state was the expertise of chemists in detecting adulterations of waters, alcohols, and tobacco for public health and taxation purposes. Their contribution to the improvement of public health included spectacular achievements, such as acid fumigations invented by Louis-Bernard Guyton de Morveau of Dijon to combat putrid emanations from cemeteries (Leroux 2017). With the emergence of pneumatic chemistry in the last decades of the century, chemists contributed to improve the quality of air, as witnessed for example in Priestley's invention of eudiometry, which used a volumetric method for testing the "goodness" (breathability) of atmospheric air, a practice which spread elsewhere in Europe, particularly in Italy (Schaffer 1990). Lavoisier's and Jean-Baptiste Meunier's collaboration for preparing inflammable air for the purpose of ballooning was another spectacular contribution of late eighteenth-century chemistry at the intersection between the advancement of knowledge and technological innovation (Thébaud-Sorger 2009).

As chemists coupled cognitive ambitions with social and economic interests they increasingly became part of governance systems (Roberts and Werret 2017). Macquer, Antoine Fourcroy, and Jean-Antoine Chaptal were members of the Bureau du Commerce (trade office) and encouraged the establishment of acid factories in a number of localities, including in close proximity of large cities. During the French revolution Guyton de Morveau assumed political responsibilities as president of the Comité de Salut Public (Public Safety Committee), and along with other chemists he contributed to the war effort (Gillispie 2004). Vicenzo Dandolo, a chemist and count from Venice, fought to save his homeland when Napoleon's army was approaching, and he later became governor of Dalmatia. These high positions of chemists in governments show that chemistry was a model of the culture of science for public welfare. Utilitarian science was highly praised at the Paris Academy of Sciences as well as at the Royal Society of London in the mid-eighteenth century. In those prestigious institutions, public usefulness was a major criterion to evaluate the quality of contributions to the advancement of science (Licoppe 1996).

Chemists actively promoted the alliance of knowing and making, between science and artisanal activities. They used it for exalting chemistry on the academic stage, either through the division between theoretical and practical chemistry, in Macquer's textbooks for instance, or through the introduction of the divide between *chemia pura* and *chemia applicata* introduced by Johan Gottshalk Wallerius at the University of Uppsala, a major figure of chemical expertise in Scandinavian mineralogy (Meinel 1983). The conceptual dichotomy between pure and applied chemistry astutely turned the proximity of artisanal activity and chemical science into a dependence of the former upon the latter; it reversed the chronological antecedence of chemical arts before chemical science into a logical consequence of arts upon science, a dependence of chemical practice upon a core theory. It provided a key argument for the promotion of chemistry in the lecture hall.

In mid-century France, however, an alternative option prevailed in the entry "Chemistry" of Diderot's *Encyclopédie*. The author of the entry, Gabriel-François Venel, celebrated the heroic figure of the *artiste* as a learned artisan contrasted both with speculative philosophers and craftsmen locked into a blind routine. Venel claimed that training a professional chemist – chemistry as a "métier" – requires lengthy and arduous experience to become acquainted with the specific qualities and quirks of chemical substances. Far from considering chemical arts as mere applications of a "pure chemistry," Venel claimed that a true chemist is first and foremost an *artiste*. He thus stressed the artisanal dimension of chemical practice including skills of glassblowing, luting, repairing, and adapting vessels. However, there is more in the portrait of the *artiste* than extensive practical knowledge and skill. The *artiste* has true knowledge acquired through manual work in the laboratory. It includes tacit knowledge, an intimate acquaintance with the behavior of chemical substances resulting from sustained repeated practice, together with systematic knowledge acquired through chemistry courses or books. The former knowledge requires the full engagement of the body, especially of the senses (Roberts 1993; Riskin 2002), and Venel (1753) considered the senses as more reliable than physical instruments:

> The learned artisan [artiste] we speak about will never think it sufficient to judge the degree of the heat he uses by thermometers, or the number of drops during a distillation process by a clock with a second hand. Rather, as technicians often say, he will have his thermometer in the tips of his fingers and his clock in his head.

Venel's portrait of the chemist-artiste embodies Diderot's ideal of experimental philosophy: chemists had a hybrid social status between the "philosophers who only have ideas" and the operators (*maneuvres*) who endlessly proceed with no specific cognitive purpose (Diderot 1753). Both Diderot and Venel thus praised chemistry for being able to address the concerns of both artisans and savants: "Chemistry contains within itself two different languages, one popular and the other scientific" (Venel 1753). If we assume, alongside sociologists and linguists, that language is the marker *par excellence* of social classes, then chemistry could be characterized by its subversion of the social distinctions between groups. It rather distinguished itself as a social practice oriented toward public utility and the common good.

As chemists were investigating the three realms of nature – minerals, plants, and animals – they were at the core of all efforts to stimulate agricultural, industrial, and social improvement. In particular, they were key players in the development of cameralism (*Kameralismus*), a German movement aimed at increasing the efficiency of employees in the state administration (Tribe 2005). In order to train skilled officials and to advise the managers of institutions, a number of German universities developed courses in police (*Polizei*) and oeconomy in the second half of the eighteenth century (Wakefield 2010). For instance, Johan Beckman (who coined the term *Technologie*) was appointed as ordinary professor of oeconomy at the University of Göttingen in 1770. The eighteenth-century disciplines police and oeconomy have little connection with our modern concepts of police and economics. Based on the vision of humans as integral parts of the natural order, the central idea of the cameralist oeconomic science was that human activities had to be integrated with natural mechanisms and cycles (Schabas 2005). Therefore, cameralism contributed to the promotion of chemistry as a utilitarian approach to nature in German universities (Hufbauer 1982; Wakefield 2010). In this perspective chemistry was reconceptualized as a hybrid science linking social welfare with the management of natural resources. The notion of *oeconomy* forged from the Greek term *oikos* (home) also linked the public and the private spheres. It referred to a realm of activities

> in which the investigation of nature merged seamlessly with concerns for material and moral well-being, in which the interdependence of urban and rural productivity was appreciated and stewarded, in which "improvement" was simultaneously directed toward increasing the yields of agriculture, manufacturing and social responsibility.
>
> (Roberts 2014: 134)

The analogy between household and chemistry lies in their management of materials. Werrett (2017) argues that chemistry contributed to extend the management of domestic space to urban areas, to the state, and to the world. He characterizes oeconomic management by its frugality, care, maintenance, and repairs of materials. Balancing the inputs and outputs, transforming, recycling, this "prudent stewardship of materials" was a moral attitude as much as a way of saving materials and money. The material culture of chemistry was a well-balanced order involving not only measurement practices but also a correct moral balance. Once again, Lavoisier appears as an heir to this material culture rather than as the initiator of new practices. His various uses of the balance as an instrument, an accounting practice, and a general principal of husbandry are exemplars of the chemists' promotion of oeconomy. Furthermore, Lavoisier's multiple careers as an academic scientist, a tax collector, an estate manager, an economist, and a social reformer can be viewed as an epitome of the *oeconomic* balance between the social and the natural worlds (Bensaude-Vincent 1992).

CHEMISTRY ON THE ENCYCLOPEDIC MAP OF KNOWLEDGE

The *Encyclopédie* is a crucial source for clarifying the relations of chemistry with neighboring sciences and its place in natural philosophy, because its aim was precisely to display a metaphorical map of knowledge. "It is a kind of world map which is to show the principal countries, their position and their mutual dependence, the road that leads directly from one to the other" (D'Alembert 1751). The encyclopedia's "Preliminary Discourse" presented a systematized chart derived from Francis Bacon's tree of knowledge. Based on three mental faculties, it consisted of three branches: History, the practice of memory; Philosophy, the practice of reason; and Fine Arts, the practice of imagination. Philosophy included four divisions: metaphysics, theology, humanities, and natural sciences. The latter branch was divided into Mathematics and Physics. The former encompassed three subdivisions – pure mathematics, mixed mathematics, and mathematical physics – while Physics encompassed astronomy, meteorology, cosmology, mineralogy, botany, zoology, and chemistry. In the mid-eighteenth century, Physics was a generic term referring to the study of nature, rather than to a disciplinary field dedicated to a specific class of phenomena such as heat, electricity, and magnetism. Thus, in the tree-like map outlined in the "Preliminary Discourse," chemistry featured as a modest part of Physics, at the bottom of all natural sciences, next to "natural magic" and superstition.

The entry "chemistry" in volume 3 completely subverted this hierarchical organization. Venel developed a lengthy defense of chemistry in response to Fontenelle's description of the "spirit of physics" as "cleaner, neater and less encumbered" than the "spirit of chemistry," owing to the clarity of Cartesian mechanics. Venel made the case for the recognition of chemistry as a branch of natural philosophy on a par with mathematical physics, using three major arguments. First, he criticised the alleged superiority of mechanical interpretations of chemical reactions, which he presented as the illusory reduction of individual qualities to geometrical properties. Using Stahl's distinction between "aggregate" and "mixt," he argued that all phenomena related to aggregative unions were ruled by mechanical theories, while the composition of mixts could only be explained in chemical terms. Second, he promoted chemistry as an alternative science that takes up the challenge of understanding obscure qualities, whereas physics denied their existence; he declared that Stahl's *Specimen beccherianum* (1703), the acme of "chemical genius," was the exact equivalent of Newton's *Principia* (1687). Third, not content with this attempt to put chemistry on the same footing as mathematical physics, Venel went even further and credited chemistry with the dignity of wisdom. He distinguished the *"sapientia chimica"* as the capacity to resist the illusory power of generalizations.

> A few half-philosophers might be tempted to believe that we are aiming at high generalizations. But we affirm on the contrary that we have clung to those notions that spring the most directly from concrete facts and knowledge and that can help us understand Chemistry's practical side.
> (Venel 1753)

Thus, the *Encyclopédie* provided a stage for contrasting the identities of physics and chemistry. Newtonianism was redefined as a deductive mathematical approach to matter, similar to Cartesianism, and contrasted with the empirical and practical approach of chemistry. Henceforth it was assumed that Stahlianism (see Chapter 1 in this volume) was a chemical approach, and Newtonianism was a physical approach.

CHEMISTRY BEHIND THE FACADE

Although it was a prestigious showcase, the *Encyclopédie* may provide a biased view of the actual organization of knowledge when taken at face value. The territories delineated and the crosslinks created do not give us a faithful picture of the many and varied actual interactions between natural scientists, which very much depended on local circumstances and opportunities. Furthermore, the *Encyclopédie* provides a static view, which was soon considered obsolete. The publishers' plan to update the *Encyclopédie* in the 1780s ended up overturning the mid-century's map of knowledge. The publishers gave up the alphabetical order of the first edition in favor of a series of disciplinary dictionaries. The place of chemistry on the map of knowledge was therefore reassessed according to multiple geographical and historical circumstances. The uniform facade of the Enlightenment had to be shattered on account of local audiences and situations.

In Scotland, for instance, chemists such as Cullen and Joseph Black enthusiastically supported a Newtonian view of chemical affinities, and trained generations of chemists to adopt a corpuscularian view of matter (Anderson 2015). In France, Venel himself presented a more nuanced exposition of chemistry – quite different from the heroic image of the anti-Newtonian science he had drawn earlier – in the courses he delivered in Montpellier in the 1760s (Lehman 2009a). In the French provincial academies the profile of chemistry was adapted to local circumstances. As chemists supported local entrepreneurs, they created various alliances of chemistry with natural history, industry, or agriculture, depending on the local public, the local professional guilds, or the presence in France of manufacturing enclaves protected by royal *privilège* (e.g. Rouen).

There was certainly no uniform culture of chemistry across Europe (Bensaude-Vincent and Abbri 1995). In Sweden, chemistry developed in connection with mining, metallurgy, and mineralogy, and was particularly instrument-oriented. In the Netherlands, the culture of chemistry was intimately linked to studies of electricity and medicine; in the Belgian provinces, to medicine and metallurgy; but in Spain, chemistry developed in connection with industrial, medical, and military projects (Serrano 2017). In Northern Italy, physiologists developed a strong interest in the chemical study of spa and mineral waters, while Vicenzo Dandolo contributed to wine making and agriculture. In the 1770s other Italian chemists, including Marsilio Landriani, Alessandro Volta, Felice Fontana, and Giovanni Fabbroni, were at the forefront of pneumatic research (Abbri 1991).

This heterogeneous landscape should not suggest that these cultures of chemistry were isolated from each other. On the contrary, the enlightened practitioners of chemistry developed strong international connections, either through institutional channels (especially memberships in foreign academies), or through individual initiatives of correspondence and travel. Chemical views and news traveled between the great capitals of Europe, and even to peripheral sites in Philadelphia or Mexico. Guyton de Morveau played an active part in these discussions through his correspondence (Grison et al. 1995). Among the most active networkers, the Portuguese Jean-Hyacinthe de Magellan not only corresponded but also traveled between Paris, London, and Leuven. He thus introduced Priestley's ideas to Lavoisier (Malaquias 2008). This dense international network proved extremely helpful for the circulation of the new language of chemistry, and for Lavoisier's oxygen theory in the last decade of the century (Bensaude-Vincent and Abbri 1995).

Concerning the boundaries of chemistry with neighboring fields, the entry "chemistry" in the *Encyclopédie* is also somewhat misleading. Although Venel's plea for the philosophical dignity of chemistry involved the dissociation of the public image of chemistry from the obscure practices of pharmacists and perfumers, apothecaries could still claim to contribute to the advancement of chemistry in the 1770s. For instance, Antoine Baumé, a member of the chemistry section of the Paris Academy of Sciences, published an important textbook entitled *Experimental and Rational Chemistry* (1773), which developed significant theoretical considerations while describing practical recipes in a separate volume (Simon 2014). In Paris, pharmacy reached a higher social and institutional prestige with the creation of the College of Pharmacy in 1777. This college broke the association of pharmacy with the guild of spicers and perfumers, and instead reinforced its association with chemistry. Although this prestigious college did not survive the fundamental institutional reconfiguration during the revolutionary period, pharmacy regained its social prestige in 1803 with the creation of a national School of Pharmacy, which included much teaching of chemistry (Simon 2005).

By the end of the century the clear boundary between chemistry and physics outlined by Venel was no longer viable. A specific branch of natural sciences named "physics" gradually came into being in the second half of the eighteenth century when demonstrations of electricity and mesmerism became extremely fashionable. Physics was officially recognized as a class of the Paris Academy of Sciences in 1785, but this recognition did not contribute to demarcating the field of chemistry. From a theoretical perspective heat, electricity, and magnetism belonged to chemistry, because all three of these entities were considered to be imponderable substances or fluids (Heilbron 1993). Looking at the investigative practices of famous savants such as Joseph Black or Joseph Priestley, for instance, it is difficult to say whether they were physicists or chemists. They equally contributed to the study of heat, of electricity, and of gases. Pneumatic science in particular was a flourishing research field at the intersection of physics, chemistry, and medicine, and helped to reconfigure these territories on the map of knowledge.

Another reason why Venel's anti-Newtonian picture of chemistry in the *Encyclopédie* has to be qualified is that it did not do justice to the increasing influence of a program aimed at reconciling chemical affinities and the uniform gravitational force. The French naturalist Georges-Louis Leclerc, Comte de Buffon, considered that chemistry would be a true science only when it would be modeled on Newton's astronomy. In his *Histoire naturelle*, Buffon encouraged chemists to subsume the tables of affinities into one single law that could be expressed in mathematical form. He regarded the laws of affinity as an expression of Newton's universal gravitation. He presented the theory of gravity as the key that would unlock the mysteries of affinity, provided Newton's law of gravitation be slightly modified to take into account the shapes of the constituent particles which played a crucial role in short-distance attraction (Buffon 1765: vol. 13, 39).

Guyton de Morveau enthusiastically supported Buffon's program in the 1770s, and Macquer was attracted by the perspective of subsuming the various aspects of chemical affinities under a single force. He thus insisted that the affinity of aggregation and the affinity of composition were various effects of the same cause. In the entry "Gravity" of his *Dictionary*, Macquer suggested that the force of attraction was modulated not only by the distance and the shape of the constituent particles, but also by a variety of parameters such as the density of the particles and the nature of the contact between them. Cullen believed that the micro-force structure of the Newtonian aether, which he identified with the matter of heat and its repulsive force, offered a potential explanation of elective attractions. Furthermore, he thought that this concept was capable of being measured experimentally, by using the increase (or decrease) of the temperature during a chemical reaction as an indirect way of quantifying absorption or release of heat by that reaction (Christie 1981).

At the turn of the nineteenth century, Claude-Louis Berthollet found an alternative way to make chemistry a Newtonian science by using mechanics as his model rather than astronomy. In his *Essai de statique chimique* (1803), Berthollet challenged the chemistry of affinities when he stated that the direction of chemical reactions was not exclusively determined by the elective chemical affinities of the reagents, but should normally tend toward an equilibrium state unless a product of the reaction were removed from the reaction, for instance by precipitating out as a solid (Holmes 1962; Grapi 2001). Because in actual fact most reactions were complete, the Newtonian dream of a rational chemistry turned into a nightmare, essentially divorcing theory and practice. Berthollet's attempt to subordinate chemistry to mechanics was dropped in the early decades of the nineteenth century, but Newtonian theory, with its logical mathematical structure and causal explanations based on mechanical forces, remained an unachievable ideal for chemistry throughout the nineteenth century. In the late eighteenth century, Newton's forces of attraction and repulsion without his notion of material corpuscles inspired the Croatian theologian Ruggiero Boscovich, who assumed that the gravitational force could rule the chemical affinities because it could be either attractive or repulsive at short molecular distances. His dynamic view of matter became a source of inspiration for chemical philosophers such as Humphry Davy.

A CHEMICAL FOOTPRINT IN THE PHILOSOPHICAL LANDSCAPE

Chemistry was more than a major component of natural philosophy, it also permeated metaphysics, political philosophy, and the philosophies of language and cognition. As there was no strict boundary between science and philosophy in eighteenth-century culture, all philosophers were more or less acquainted with the sciences of their time. Indeed some of them, for instance Thomas Reid, Adam Ferguson, and Dugald Stewart in Scotland, were practitioners or lecturers in mathematics, physics, or chemistry. Chemistry proved to be a source of inspiration for inventing new doctrines and concepts, while the concepts of affinity, mixtion, and analysis easily moved through various intellectual cultures.

It would be inadequate to describe this circulation of concepts in terms of import or export between chemistry and philosophy. On one hand, there was nothing like a one-way traffic from chemistry to philosophy. For instance, language was a key concern throughout the eighteenth century. The ambition to construct a universal and rational language was shared by mathematicians, such as Leibniz, by botanists such as Carl Linnaeus and by chemists, who regularly complained about imperfect names of substances. Lavoisier explicitly borrowed the guiding principles of the reform of chemical nomenclature in 1787 from Etienne Bonnot de Condillac's *Logic*, while other chemists were more inspired by Locke's view of language as a convention (Beretta 1993; Bensaude-Vincent 2010). The reform of the chemical nomenclature was an integral part of a broader historical dynamic, which involved metaphysics, logic, and natural philosophy. On the other hand, as noticed above, chemistry enjoyed cultural prestige. Chemical notions percolated through everyday language and philosophical endeavors. Quite naturally chemical concepts came to the philosophers' minds as they wrote their essays. It thus seems more appropriate to use the metaphor of a chemical footprint on eighteenth-century philosophy.

An exemplar of this chemical footprint, albeit unique in its genre, was George Berkeley's volume entitled *Siris: A Chain of Philosophical Reflexions and Inquiries concerning the Virtues of Tar Water, and Diverse other Subjects Connected Together and Arising One from Another*, published in 1747. Berkeley, a bishop and an enlightened man of science, instrumentalized chemistry to fight atheism. This apologetic essay starts from tar and tar water, and leads up to God through a long chain of arguments. Taking inspiration from Homberg's concept of principles, Berkeley described light/fire as an active and all-pervading element which animates the entire world. This light/fire is an instrument of God (Airaksinen 2010; Petterschmidt 2010).was not attached to a specific philosophical attitude. Rather than attempting to provide a global survey of chemistry-inspired philosophies, a few examples will suggest the diversity of philosophical positions inspired by chemical concepts. While Berkeley developed chemical arguments in favor of idealism, Diderot found in chemistry inspiration for developing a materialist worldview. Diderot, who as we have noted had attended the demonstrations of chemistry delivered by Rouelle at the Jardin du Roi, considered Stahl a great chemist and a poor physician. Like Venel, he admired Stahl for his distinction between aggregation and mixtion, and like Leibniz he rejected his view of living organisms being ruled and maintained in life by an external agent, which was an immaterial vital force. Diderot developed a dynamic view of matter with immanent agencies. All bodies are made of dynamic minimal units endowed with the potential of interacting with others, according to the laws of affinity. Those active principles constituted a continuity between inert matter and living matter, between mineral stones and animals, as shown in *d'Alembert's Dream* (*Le Rêve de D'Alembert*). Central to Diderot's materialism was the notion of immanent operations, of invisible chemical operations between material elements, which led to a global view of nature as a continuous process, a *natura naturans* (Pépin 2013).

Rousseau developed the image of nature as a theater of operations, literally an opera, in an unfinished and unpublished textbook of chemistry. In his *Institutions chymiques*, presumably written in 1747–1748, Rousseau shifted the commonplace metaphor of the theater of nature toward a "laboratory of nature," in which the four elements – earth, fire, air, and water – featured as instruments of nature, a concept he found in Boerhaave's and Rouelle's courses. Rousseau rejected both Fontenelle's injunction to move behind the stage of the theater to investigate the machinery and Pluche's invitation to just enjoy the performance on the stage. He assumed that the best way to understand nature was through chemistry.[4] Although he assumed that our knowledge of nature was limited, that the ultimate particles of matter would forever remain out of reach, in order to penetrate the secrets of nature, one has to experiment the operations of nature: digestions, dissolutions, filtrations, fermentations, and calcinations. In other words, one has to perform chemical manipulations in a laboratory:

> To establish an artificial laboratory on the model of the laboratory of nature, it is not enough to take a glimpse on the pathways she uses; we must above all perfectly know the instruments she uses.
>
> (Rousseau 1999: 63)

This claim is at odds with the romantic portrait of Rousseau engaged in a direct and emotional contact with nature conveyed by his *Rêveries d'un promeneur solitaire*. It also challenges the standard view of Rousseau, derived from the first Discourse, as morally and politically contemptuous of sciences and arts. Rousseau's practice of chemistry had a significant impact on his political philosophy. In the preface of his *Discours sur l'origine et le fondement de l'inégalité*, which is its main pillar, Rousseau presented his work as a thought experiment aimed at reconstructing the history of mankind analogous to the real experiments that could be conducted in the laboratory of nature. Rousseau thus created his own method with reference to the experimental method that he praised. Moreover, Rousseau found in chemistry a source of inspiration to develop a new conception of the relationship between individual beings and their common social existence, as Bruno Bernardi (2003) has convincingly argued. Rousseau managed to move beyond two rival models in political philosophy: the biological metaphor of society as an organism, and the mechanistic model of society as a combination of forces advocated by Thomas Hobbes in *Leviathan*. Rousseau rephrased the political issue of the social contract in terms that were reminiscent of the chemical concept of mixt as distinct from aggregate. Free individual wills assembled to form a single will are like the individual principles combined into a mixt. In both cases a new unity – endowed with properties differing from those of the individual entities which make it up – is produced without intervention of a transcending principle.

Chemistry also provided philosophers with epistemological models. For instance, in his investigation of human nature, David Hume followed an experimental method that was influenced much more by the qualitative methods of chemists than by Newton's experimental method (Demeter 2012). Chemistry was even more stimulating in German universities where natural sciences were taught in the faculties of philosophy. Although in the *Metaphysical Foundations of Natural Sciences* (1786) Immanuel Kant presented chemistry as an improper science because its principles could not be derived a priori, in the second Preface of the *Critique of Pure Reason*[5] he nevertheless praised Stahl's transformation of metals into calces, and of calces into metals, as one of the founding events of modern science. More importantly, Kant found in chemistry a source of inspiration for his transcendental approach to cognition and practice (Leqan 2000). Kant's concept of analysis which he developed in the section "Transcendental Logic" of the *Critique of Pure Reason* admittedly comes from logic with influence from mathematics. However, in the section "Doctrine of elements," Kant explicitly referred to the elements resulting from a chemical analysis to define the regulative ideas of reason: they are not revealed by nature, and can be used as instruments to question nature. The union of two different substances under the action of attraction and repulsion provided a metaphor to describe the mental process of cognition through the establishment of relationships between heterogeneous elements. Again, in the *Critique of Practical Reason* (1788) Kant decomposed moral judgments into their elements, and in the *Opus posthumum*, he explicitly stated his preference for empirical chemical analysis, because it decomposed matter into qualitative pure elements, whereas the mathematical method, which resulted in quantitative atoms, was not adequate for his transcendental analysis. Besides the chemical method of analysis, Kant also adopted the dynamical view of matter inspired by Boscovich, whose vision of nature as ruled by two opposite forces – attraction and repulsion – was extremely popular in the German universities at the end of the eighteenth century.

This dynamical metaphysics provided the basis of the German *Naturphilosophie* tradition, which promoted chemistry as the most fundamental science. Although *Naturphilosophie* is usually associated with nineteenth-century Romanticism, it was rooted in the dynamical view of nature developed in the late eighteenth century. The term *Naturphilosophie* was coined in 1797 by Friedrich Schelling in an essay entitled *Ideen zu einer Philosophie der Natur* (*Ideas concerning a Philosophy of Nature*). This emerging tradition was radically opposed to the mechanical philosophy, and exemplifies the complexity of the cultural landscape in the last decade of the eighteenth century. Schelling rejected the Newtonian dream of a mathematical treatment of affinities developed in the same period by Berthollet and Pierre-Simon de Laplace at the Société d'Arcueil. He assumed that chemical forces were distinct from mechanical ones and considered all chemical reactions as a process of interaction between two polar forces.[6] Thus Schelling's *Naturphilosophie* promoted chemistry as a universal model for understanding nature, as David Knight has rightly pointed out:

> [He] made it a subject of general concern, something that the educated person ought to know about. ... Whereas previously it had been a technical subject, it became a part of culture. It was indeed presented as a dynamical and therefore fundamental science, not merely as an independent one with its own limited sphere escaping for a time from the apron strings of physics.
> (1992: 65)

Schelling and then Hegel thus increased the cultural prestige of chemistry at the very moment when the success of Lavoisier's revolution destroyed the chemical tradition, which was their source of inspiration. In this perspective the chemical revolution looks like the swan song of the golden age of the chemical culture promoted during the eighteenth century, as well as being the foundation of modern chemistry. Lavoisier's efforts to emancipate chemistry from all metaphysical views of matter in his reconceptualization of elements as provisionally undecomposed (elemental) substances and his rejection of affinities in his *Traité élémentaire de chimie* (1789) discredited the chemical philosophy promoted by Schelling. With John Dalton's *New System of Chemical Philosophy* (1808), chemical philosophy thereafter simply became just a branch of natural philosophy, no longer an attempt at understanding nature through the lens of chemistry.

CONCLUSION

This chapter has argued that the previously accepted view of eighteenth-century chemistry as an obscure empirical study simply waiting for Lavoisier's revolution is a significant misunderstanding that cannot survive a survey of the works of chemists in the context of their time (Holmes 1989). In terms of cultural importance and prestige, the eighteenth century may rightly be considered the golden age of chemistry. With its profile of experimental philosophy challenging the dogmatic "spirit of system," a science intimately connected to daily life, moreover in the service of public welfare and a strategic asset for the prosperity of nations, chemistry appeared as a glamorous endeavor for all educated persons. This does not, however, mean that chemistry provided a unique and uniform overarching model. There were a variety of chemical cultures accommodated to local or national contexts. The uniform facade of Enlightenment as seen in the pages of the *Encyclopédie* should therefore be replaced by the image of a mosaic, assembled with variously colored and patterned materials. These variations, accountable in terms of the geographies, histories, and interests of the local sites where chemistry was practiced, nevertheless make up a coherent global picture. The activities of the practitioners and their audiences contributed to the rapid pace of chemical development in the second half of the century, when it featured as the fundamental science embracing a wide range of natural phenomena such as affinities, heat, electricity, and magnetism. In the last decade of the century, however, the culture of chemistry was fractured into a variety of rival and competing research schools that no longer constituted a unified coherent picture.

CHAPTER FIVE

Society and Environment: *Chemistry and Daily Life during the Eighteenth Century*

MATTHEW DANIEL EDDY

INTRODUCTION

On a wintry day in 1780 the first meeting of the Baptist's Head Coffee House Philosophical Society was called to order. Its members included some of London's most eminent experimentalists. Over the next seven years the club would meet to discuss a wide range of topics, many of which related to cutting-edge chemical discoveries in medicine, natural history, and industry. The members were eminently sociable, discussing the most recent news communicated to them from across Europe and its colonies. They treated chemistry as a bold new science that was directly relevant to society and the economy. They discussed topics relevant to various forms of patronage and, although they avoided political topics, their views of chemistry were shaped by ideological commitments (Levere and Turner 2002). The members of the Society also demonstrated a growing awareness of the ways in which the processes and cycles of the material world could be used to understand the environment and ecology. Yet, although the biographies of the Society's members were impressive, their scientific interests were not unique to London. Indeed, from Berlin to Bombay chemistry was discussed in clubs, shops, surgeries, parlors, and even kitchens. Whether in Amsterdam or Aleppo, chemistry became increasingly relevant to society and the environment over the course of the long eighteenth century. This chapter charts this phenomenon and emphasizes its importance to social, cultural, and environmental history.

PATRONAGE

Chemistry became a leading science during the long eighteenth century. It was taught in a variety of institutions. It was ripe with discovery. It inspired sophisticated instruments. Nevertheless, no matter how stimulating it might have been, its practitioners still needed to earn a stable, sustainable income and, like most people, they longed for social status. How did they achieve these aims? The truth is that chemistry operated across many trades and professions, fostering the emergence of hybrid experts, that is, specialists able to master and adapt chemical knowledge across many commercial domains (Klein 2017). In this context experts were practitioners who learned about the science of materials at university or through an apprenticeship.

Chemistry served as the basis for a source of employment in the rising middle class. Metallurgists assayed minerals in mines. Apothecaries combined substances into drugs. Physicians employed chemical theories to understand health and disease. Artisans skilfully transformed the materials of nature into commodities. In many respects, chemists were often their own patrons, funding experiments out of their own pockets. Sometimes these experiments helped chemists to invent trade secrets that were lucrative, leading journalists and artists to depict them as being greedy or unscrupulous servants of unseemly patrons (Figure 5.1). In many cases chemistry was so integrated into some occupations that its practitioners did not readily describe themselves as chemists. The scale of the substances and instrumentation required for occupational chemistry were modest. The home or shop provided laboratory space. Spouses, children, and servants were technicians. Everyday household items provided the material basis for experiments (Werrett 2017).

There were many opportunities for entrepreneurial chemists whose personal connections or sheer persistence generated occasional opportunities for financial remuneration. Although now recognized for his contributions to biology and systematics, the Swedish physician Carl Linnaeus, for instance, gave lectures on the chemistry of mining in the Swedish Assay Office during his early years to supplement his meagre income (Fries 1923: 106). In Scotland, the Reverend Dr. William Laing used household instruments to determine the

FIGURE 5.1 William Hogarth, *In the Cabinet of the Quack Doctor, the Viscount Squanderfield Holds Out a Small Pill-Box as a Young Girl Dabs Her Face with a Handkerchief*, colored aquatint after William Hogarth. Published in France and based on William Hogarth's original (London: 1745). The print was Plate III in Hogarth's series entitled *Marriage-a-la-Mode*. Courtesy Wellcome Collection CC BY.

chemical composition of spa water, an act that led the University of Aberdeen to grant him a medical doctorate that allowed him to earn money as a practicing physician (Eddy 2010). In France, savants, artisans, and *artistes* used chemistry to win prizes offered by scientific societies and academies, a move that elevated their status as experts and acted as a form of advertising (Bertucci 2017).

Stories of entrepreneurial chemists abound, but this kind of activity usually did not provide them with the kind of patronage required for expensive or ambitious chemical projects. The time, space, and resources required to pursue large-scale research and development projects required a patron, that is, a person or an institution providing significant financial assistance. Three notable categories of patrons throughout the century that provided this kind of support were the state, nobles, and industry.

National and local forms of government consistently employed chemically trained experts to improve or regulate the substances and compounds used to make munitions, coinage, food, paper, and drugs. The European state, which often included a cooperative effort between the monarch and the legislature, aided the spread of chemical knowledge by funding university professorships. In Britain, King George III established the Regius Chair of Natural History at the University of Edinburgh in 1779, which made it possible for chemical mineralogy to be taught in its medical school to aspiring industrialists, physicians, and entrepreneurs (Eddy 2008). In Spain, the government created several chemistry chairs during the late eighteenth century, moving all of them into a state-sponsored laboratory in Madrid called the *Real Escula de Quimica* in 1799 (Bertomeu-Sánchez and García-Belmar 2000).

Some nobles saw chemistry as an intellectual hobby or as a mode of increasing the productivity of their land, but their support was crucial for many chemists. Antoine Lavoisier, a member of the French nobility, used his family wealth to finance the expensive and exquisite precision instruments that provided evidence for his new ideas about chemical substances (Beretta 2014a). Aristocrats offered patronage through employing chemists and financing their research. Here women played a notable role. The Duchess of Devonshire, for instance, was so fascinated by experimentation that she employed chemists to construct a laboratory in one of her houses. She also supported Thomas Beddoes' plans for his Pneumatic Institution in Bristol, allowing him to eventually pursue his widely publicized experiments on the newly discovered gas that we know today as laughing gas or nitrous oxide (Jay 2009: 96, 101–2).

The ultimate aristocratic patrons were monarchs, especially those seeking to improve public health, the military, or coinage. Such was the case for the chemists under the patronage of Peter the Great of Russia. When St. Petersburg became the capital in 1713 he appointed the chemist Jacob Bruce to oversee the *Glavnaia Apoteka*, the Chief Pharmacy, a massive medical chemistry complex with laboratories that provided drugs for the nobility and the military. In later years, Peter placed Bruce in charge of the Russian artillery, the Board of Mining and St. Petersburg's mint, all of which had crown-funded assaying laboratories designed to test the purity and strength of metals. The scale of the research conducted by Bruce and the many students and mining officers in the laboratories was enormous and placed Russia on a firm footing when the Ural Mountains and Siberia became a global supplier of metals later in the century (Werrett 2013).

Near the middle of the century, the Industrial Revolution instituted a new paradigm in scientific patronage, particularly for chemical research and development in places like the English Midlands, Belgium, and Silesia. Chemists were needed to analyze the purity or composition of an increasing number of industrial substances such as dyes for fabrics, bleach for linen, and fuels for furnaces. Large companies, such as that of Birmingham's Matthew Boulton and James Watt, employed whole teams of chemical experts to design new steam engines and gas instruments, and explore the by-products of combustion. Coal alone yielded a number of saleable products such as coke, oil, tar, pitch, and "inflammable air" (hydrogen). By the end of the century, the funding provided by factories and manufacturers began to rival that of governments and aristocrats.

SOCIABILITY

During the seventeenth century the public perception of chemistry was influenced by romanticized or sensationalized depictions of alchemists and artisans. Through the publications of authors such as Robert Boyle in Britain and Georg Ernst Stahl in Prussia, the Republic of Letters and members of polite society increasingly viewed chemists less as "sooty empirics" and more as experts who possessed sophisticated material knowledge. This transformed the view of chemistry into a science that was directly relevant to the forms of sociability that blossomed during the long eighteenth century.

Within polite society it was the responsibility of ladies and gentlemen to be urbane conversationalists. Chemistry proved to be a stimulating and captivating topic, providing delectable morsels of polite stimulation at dinner parties and in correspondence. It also supplied useful and impressive factoids about the new industries that favored economic stability of Europe via the production of commodities. To facilitate the acquisition of chemical knowledge, some affluent parents introduced chemistry into the education of their children. When Prince Pavel Michailovich Dashkov, the son of Princess Ekaterina Romanovna Dashkova of Russia, went to study in Edinburgh during the 1770s, his mother wrote a letter to the university's principal William Robertson that identified "the first principles of chemistry" as a subject she wanted her son to learn (Dashkova 1840: 122).

For those who did not have the time or inclination to read books about chemistry, popularizers traveled across Europe and its colonies, offering courses on chemical topics that lasted for a day, several weeks, or even an entire year. Many such lecturers attracted a notable number of female attendees. The Philadelphia author and lawyer Francis Hopkinson once noted that the 1785 American lectures of the blind chemist Henry Moyes attracted over 1,200 people and were "very popular and a great favourite of the Ladies" (Anderson 2017: 101). Attendees had to be wary, because lecturers sometimes had ulterior motives. The lectures offered by Paris' Jacques Bianchi and London's John Theophilus Desaguliers, for example, featured sparks and bangs that acted as exhilarating advertisements for the specialty instruments that they sold (Desaguliers 1717; Golinski 1992; Lynn 2006: 25–7).

The presence of women in public lectures reminds us that chemical knowledge played a role in domestic settings. In an age when chemistry was often pursued in people's homes, there were opportunities for women and girls to learn chemistry, or to be involved in either observing experiments or even participating in them (Leigh and Rocke 2016). Glimpses of this role sometimes appear in visual culture. A 1773 allegorical engraving of Charles-Nicolas Cochin the Younger's drawing "La Chimie," for instance, pictured a woman conducting experiments with a furnace in a room bearing the features of a neat, affluent kitchen (Figure 5.2). Moving from allegorical to a more literal example, Marie Lavoisier's pen-and-ink drawings of her husband's experiments depict herself as a participant (A.-L. Lavoisier 1790).

Chemical knowledge was especially present in female forms of sociability that revolved around polite conversation, culinary culture, and public health. The Duchess of Devonshire, for instance, sought information on new experiments, writing to leading scientists such as Sir Charles Blagden, the Secretary of the Royal Society, to enquire: "Has any thing new arisen in the last fortnight? – Any chemical, mineralogical or philosophical novelty?" (Wills 2019: 155). In many respects, early modern women and girls were already familiar with a variety of chemical substances because many of the ingredients used in experiments were the same as those discussed in herbals, cookery books, pharmacopoeias, and first-aid manuals.

Likewise, chemical processes like distillation, fermentation, and putrefaction were part and parcel of everyday household activities such as brewing and making preserves. Because cooking and dining were important forms of female sociability, these chemical substances featured in day-to-day discourse. In this domain, women were culinary chemists (Leong 2008; Spary 2012; Spary 2014). Sometimes women pursued chemical discourse outside the home within salons or societies. The members of Madrid's *Junta de socias de honor y mérito* and the *Asociación de señoras*, for example, used chemical ideas to discuss the relationship between health and their rapidly urbanizing society (Serrano 2013).

The desire to know more about chemistry was aided by the fact that the early modern period was the age of lexicons and dictionaries, some of which focused on chemical topics relevant to medicine and mining, and, later in the century, on agriculture and industry. An excellent case in point is Stephan Blancard's best-selling medical dictionary *Lexicon Medicum*. Written in German, Greek, and Latin, the book offered helpful definitions or descriptions of terms relevant to the practice of medicine. Blancard was an eminent doctor based in Amsterdam who, like many of the physicians educated in the Netherlands, Scotland, and German states, used chemistry to explain various cures and causes of disease. When it was republished in 1718, the eminent German experimentalist Georg Ernst Stahl, an advocate for medical chemistry, wrote a supportive preface. Chemically oriented dictionaries remained strong throughout the century.

Specialized lexicons such as Pierre Macquer's *Dictionnaire de chymie* (1766) appeared as the century progressed. Macquer was the resident chemist at the French *Académie des Sciences* and his book served as an introductory text for

FIGURE 5.2 A female figure performing chemical experiments with a furnace: representing chemistry. Etching by E.-J.-N. de Ghendt after C.-N. Cochin the younger, 1773. Courtesy Wellcome Collection CC BY.

university students and the general public alike. Dictionaries of this nature were extremely important because the language and nomenclature of chemistry was not standardized (Crosland 1962). Although Latin sometimes served as a shared scientific language for scholarly communities, much of experimental chemistry was conducted in the vernacular, with terms being drawn from the language used by artisans or miners. Reading or translating dictionaries help to codify and standardize the language and nomenclature that chemists used to express their ideas and to build experimental communities (Bret 2016).

Print culture also provided a way for chemists to be sociable with each other. Ever since the late seventeenth century, scientific organizations had been publishing journals designed to spread useful knowledge and to foster communication between their members. Perhaps the most popular publications were the *Philosophical Transactions of the Royal Society* and *Histoire de l'Académie Royale des Sciences*. These and other periodicals fostered like-minded communities of scholars and patrons. Their pages frequently featured essays and letters about chemical experiments and discoveries. Within this medium chemists debated the theories and applications of chemistry.

As the century progressed, the discipline of chemistry grew at a rapid rate and its practitioners were keen to develop communication networks that fostered the spread of new ideas. Many chemists began to call for specialized journals to meet this need. One of the scholars who answered the call was Florens Lorenz Friedrich Crell, a German physician and professor at the University of Helmstedt. Crell founded the *Chemisches Journal* in 1778. Following the lead of other Enlightenment periodicals, Crell saw the journal as a vehicle through which new forms of knowledge could be popularized in a way that contributed to the improvement of society at large. The journal went through a number of titles, but over the next two decades it served as a forum for chemistry and as a template for the establishment of other journals that had similar aims. In many respects Crell was a chemical journalist, wooing authors for information and using surreptitious appeals to German nationalism to attract patrons (Hufbauer 1982).

Crell skilfully used letters to acquire and transmit chemical knowledge. Like other periodicals of the day, his journal published letters as articles. His efforts in this area benefitted greatly from the Republic of Letters, an Enlightenment phenomenon in which like-minded thinkers shared information that improved the mind, body, and society. At a time when transportation was slow and expensive, writing letters constituted an important form of sociability. Consequently, many chemists, especially eminent ones such as Scotland's Joseph Black and Sweden's Torbern Bergman, developed their own correspondence networks to politely acquire and disseminate chemical knowledge. They wrote each other about ideas they gleaned from books, lectures, society meetings, and personal observation. They also had to learn to discern the difference between exchanging useful information and stealing trade secrets (Bergman 1965; Anderson and Jones 2012).

Chemical networks were active outside Europe as well. The brothers Alexander and Patrick Russell, both of whom served as medical doctors for the British Levant Company's base in Aleppo, Syria, are an excellent case in point. The use of "chemical medicine" had been known in the Ottoman Empire since the late seventeenth century when Ṣāliḥ ibn Naṣr ibn Sallūm, a court physician in Istanbul, translated the works of Oswald Croll, Daniel Sennert, and Herman Boerhaave into Arabic. Nevertheless, it was still difficult for the brothers to acquire specialty ingredients in Aleppo, leading them to purchase substances from company contacts in Istanbul. Similar chemical supply networks provided substances and chemical information within other European trading companies as well (Van den Boogert 2010: 145–84).

As revealed in Crell's journal and in the correspondence of other chemists, the interface between sociability, media, and chemistry significantly expanded during the eighteenth century in a way that inspired the inclusion of chemical knowledge in literary works. The sensory nature of chemistry resonated with many authors. In addition to the visuality of experiments, many of its laboratory procedures depended on the sense of smell, touch, taste, and hearing. Such empirical sensations led the mind to draw relations between natural objects in a way that influenced perception. Sensations such as sweetness and bitterness were bodily reactions to experimental substances that novelists such as Henry Fielding used to describe the personalities of characters (Thompson 2017). Tobias Smollett wrote a novel in which the narrator was an atom who commented on politics, and Johann Wolfgang von Goethe's *Elective Affinities* portrayed individuals as substances that were destined to be attracted to each other (Smollett 1769; Goethe 1999).

CONSUMERISM

Perhaps more than any other form of natural and technical knowledge, chemistry was the science that impacted the daily lives of ordinary folk. Some chemical substances were made, bought, and sold in a way that rendered them accessible not only to experimentalists, but to society at large. Those who could not afford to buy them turned to local experts or manuals that explained how to extract substances from materials found around the home and in the countryside, or in rubbish heaps and industrial wastelands. Once obtained, artisans, professionals, householders, and entrepreneurs used chemical operations to convert substances into compounds that were sold as products in shops and markets, feeding the rising consumerism that increased during the course of the century.

The circulation, production, and consumption of chemical substances and compounds was dependent upon instruments and devices which could be used to contain, measure, and manipulate matter. Experiments required a variety of flasks, pipes, filters, and valves that were made from glass, metal, and paper. The chemistry of heat depended upon the use of furnaces, pumps, bellows, burners, and smokeless chimneys. Experimental products were quantified with scales measured by instruments such as barometers, thermometers, pyrometers, hydrometers, eudiometers, and photometers. Health and safety risks to experimentalists were mitigated through the use of masks and ventilators. Those with limited resources used jars, pots, pans, spoons, stoves, and wheelbarrows, a situation that undoubtedly provided extra motivation for householders to buy items that served both domestic and scientific purposes (Holmes and Levere 2000; Golinski 2007: 108–36; Thébaud-Sorger 2018).

One of chemistry's main contributions to eighteenth-century consumerism came via the role that it played in identifying and transforming materials that were useful to the rising market economy spreading through Europe and its colonies. Nowhere was this relationship more apparent than in the world of household goods and luxury items. There were many products that were made from or with airs, earths, salts, metals, inflammables, or water. In many cases there was a reciprocal relationship between product development, consumer demand, and chemical discovery, creating a symbiotic form of technoscience (Klein and Lefèvre 2007; Klein and Spary 2010).

One of the most striking cases concerning the relationship between chemistry and consumerism is Chinese porcelain. "China" became all the rage as the century progressed. The demand for it was driven by Europe's increasing contact with India, China, and Indonesia through trade and colonialism. These areas were increasingly exoticized in art and literature, leading to a romantic desire for china and other Asian products. State-monitored trading companies such as the Dutch East India Company, the French Levant Company, and the British East India Company sought to meet the demand. Owing to the nature of early modern travel, transporting porcelain vessels from Asia to Europe without breakage was difficult and costly. Moreover, Chinese artisans would not reveal the formula of porcelain, making its materials and production a closely guarded trade secret. Consequently, for Europeans, particularly Central and Eastern Europeans, porcelain was extremely expensive. The increasing demand for affordable porcelain across Europe drove investors to fund chemical experiments aimed at finding a formula that could either replicate or rival Chinese porcelain (Berg 2005: 126–8; Gleeson 2013). The drive was especially strong in German principalities. Although they did not find the exact Chinese formula, chemists found similar ingredients that allowed them to create another kind of hard porcelain. The establishment of royal ceramics laboratories in Dresden, Meissen, and Berlin soon followed (Klein 2013; Klein 2014a; Klein 2014c).

It is in the craze for porcelain that we can learn a valuable lesson about the symbiosis between chemistry and consumerism. One of the key ingredients of Chinese porcelain was a type of fine white clay called kaolin. The demand for porcelain drove chemists to conduct experiments feverishly to find a mineral that had similar chemical properties, but their research was inhibited by a problem: hardly any systematic experimentation had been done on the chemical composition of minerals other than those important in mining. The problem was recognized in the early eighteenth century by Johann Friedrich Böttger and Ehrenfried Walther von Tschirnhaus, who conducted experiments on minerals that yielded a recipe for making true porcelain. Several decades later, Johann Heinrich Pott, a member of the Berlin Academy, extended this research by conducting over 30,000 experiments on different minerals.

Many of Pott's experiments ended in frustration. Nor was he able to crack the porcelain formulas that were used in Meissen. As is often the case in science, his failure was successful in another way, because he used his results to formulate one of the first internationally viable chemical classifications of the earths, that is, non-metallic minerals found in geological strata. Using heat and acids, he suggested that there were four basic kinds of earths contained in all rocks: calcareous, gypseous, argillaceous, and vitrifiable. He published his results in *Lithogéognosie* (1743), and they contributed significantly to chemists' and mineralogists' understanding of earths, and further laid the foundation for a new, chemical, understanding of the history of the terraqueous globe. In short, porcelain, a luxury product, drove experiments that laid the chemical foundation for modern geoscience (Laudan 1987: 49; Eddy 2008: 90–7, 107–8).

The case of porcelain is merely one example of the way in which chemical knowledge underpinned the creation or improvement of products that drove eighteenth-century consumerism. Chemical experiments and processes provided combustibles for pyrotechnics (Golinski 2017), essences for fashionable perfumes, fixed air (carbon dioxide) for tonic water (Eddy 2010), simples for drugs, spirits for liquors (Spary 2012; Spary 2014), gas for street lighting (Tomory 2012), dyes for fabrics (Nieto-Galan 2001), alkalis for bleached linen, coke for smelted steel, gums for paint, acids for burning images into copper printing plates, and inflammable air (hydrogen) for balloons and more (Kim 2017). In short, as shown in Chapter 6 of this volume, there were numerous chemical synergies between industry and trade.

Chemistry's importance to the foregoing products was acknowledged in the public sphere, particularly in popular visual culture where chemical instruments and containers were increasingly depicted in commercial contexts (see also Chapter 8 in this volume). The English artist William Hogarth, for instance, used a range of chemical and alchemical symbolism in his works. Of particular note is his *In the Cabinet of the Quack Doctor* (1745), which depicts three syphilitic patients queuing for mercury in a mid-century apothecary shop (Figure 5.1). Mercury was a chemical substance that acted as the cure for syphilis. Chemical imagery saturated the scene. The shelves are full of chemical supplies, there is a

Leyden jar on the floor, and the apothecary's laboratory in the adjoining room features an expensive collection of experimental glassware. The image clearly resonated with the public because it was reprinted several times in England and in France (Heyl 2013).

As in the case of porcelain, the demand for consumables led to chemical discoveries. The isolation of fixed air by Joseph Black, for instance, occurred in a set of experiments in which he was attempting to find a cure for bladder stones. He and other members of Edinburgh's Royal College of Physicians frequently practiced this kind of exploratory experimentation to improve their pharmacopoeia, the recipe book used by many doctors and apothecaries to make drugs (Cowan 2001). The definitive isolation of fixed air turned out to be a major discovery that affected chemical theory and practice. It showed that atmospheric air, which previously had been treated as homogeneous, was, in fact, heterogeneous.

The demand for material commodities created the need for experts who could identify, assay, and test chemical substances in the field. The search could take place in one's own country. There was a longstanding early modern tradition, for example, of savants and naturalists scavenging the countryside and seaside for animals, plants, and minerals that contained marketable substances (Cooper 2007; Cook 2016). Or the search could take place abroad, through colonial exploration and voyages of discovery, that is, journeys made possible through the networks of empire. Such was the case of the massive overland expeditions made by Simon Peter Pallas across the eastern frontiers of Russia, or of the transoceanic voyages of Britain's Captain James Cook across the South Seas. These and other European travelers were geo-prospectors and bio-prospectors whose chemically trained teams recorded and collected thousands of plants and minerals that were sent back to Europe as potential substances that could be commodified (Schiebinger 2004; Vogel 2013: 174).

Whereas ascendant empires like Russia, Britain, and France were seeking new substances across the globe, other colonial powers sought to consolidate and standardize the supply of longstanding chemically refined commodities. Here the goal was to convert substances into materials that were easier to transport and sell. Such was the case for the ores and precious metals that had been smelted and assayed in Spanish and Portuguese mines in South America. By the end of the century Spain was sending chemically trained colonial administrators to manage commodities abroad. In 1790, for instance, it sent the administrator Vincente Olmedo to Quito to co-direct the royal reserve of cinchona trees that contained quinine, a febrifuge developed by Quechua Indians of Peru that cured tropical fevers. Styled a "chemico-botanist," he was tasked with preparing a quinine extract on site so that it could be sent back to Spain for sale. A similar responsibility fell to many agents managing lucrative pharmaceutical ingredients in colonial settings (De Vos 2007; Crawford 2014).

POLITICS

The relationship between politics and society underwent significant changes during the eighteenth century. Chemistry played an important role in this transformation because of its relevance to political epistemology, that is, to the larger relationship between political beliefs and useful knowledge. At one level, it provided materials used to improve society. Although the modes of political support or guidance for industrial, military, and medical industries differed, it was acknowledged across the ideological divide in most countries that the material knowledge provided by chemistry somehow needed to be fostered (Jacob and Stewart 2004).

German cameralists, for instance, made chemistry an essential component of the educational regime used to train civil servants (Meinel 1988), a move that helped create a coterie of artisanal and savant experts who were well versed in chemistry (Klein 2012a; Klein 2012c), and who could judge the feasibility of state-sponsored projects that sought to use chemical knowledge to improve agriculture, mining, and industrial technologies. French savants were happy to support a system in which aristocrats combined their resources with state funds to create laboratories oriented toward discovering new products or the creation of purity standards that could be used in trade (Lehman 2014). Despite their commitment to *laissez-faire* economics, Britain's liberal politicians supported laws that protected chemical substances like common salt, or which regulated the material composition of drugs and food (McArthur 1801: 74–117).

Chemistry's connection to political epistemology remained strong because it also supplied metaphors that were used to characterize social interaction and, more broadly, the political establishment's relationship to the public. Prior to the eighteenth century, authors such as the ancient philosopher Lucretius and the English philosopher Thomas Hobbes advanced the notion that political theory should reflect the idea that the world only consisted of matter and forces of nature. Those inclined towards this position during the eighteenth century used experimental chemistry to provide naturalistic metaphors calculated to undermine the authority of established institutions like the church or state (Shapin and Schaffer 1985).

Using the metaphors of explosion and equilibrium, radicals and republicans argued that political systems that did not allow the needs of the public to offset the desires of the landed classes would eventually explode and rebalance themselves as representative democracies. Sometimes this kind of explosion became the literal outcome of a political clash of ideas. When Joseph Priestley, for instance, used explosive metaphors to defend the early years of the French Revolution, his Birmingham laboratory was set alight by an inebriated patriotic mob. Its fantastically explosive demise ruined him financially and he was forced to emigrate to the shores of the newly formed American republic.

Chemical explosions were an irresistible trope in the political satires created by graphic artists for public consumption. London's James Gillray, for example, published a number of entertainingly irreverent prints of this nature in the years following the French Revolution. His images thrived on political explosions. They ranged from scenes in which conservative agitators such as the politician and journalist Edmund Burke erupted from a cloud to surprise the "atheistical-revolutionist" Joseph Priestley, to an incendiary tableau of King George III as an alchemist manipulating an alembic that contained members of parliament (Gillray 1790; Gillray 1796).

The French received Gillray's attention as well. His *French Generals Retiring on Account of Their Health* (Figure 5.3) depicts the French leader Louis Marie de La Révellière-Lépeaux in an apothecary's shop in 1799, the year in which he was forced to step down from the Directorate on account of unpopular domestic policies and an unsuccessful campaign in the War of the Second Coalition. Instead of containing traditional chemical substances, the jars spread around him in this image contain various political departments, personalities, and attitudes. Despite being surrounded by renowned political and military advisors, Révellière-Lépeaux's experiment reveals that he, like an untrained

FIGURE 5.3 Larévellière-Lépeaux sits in a disordered quack doctor's room, in the presence of seven wounded French generals, one of them vomiting; representing French defeats in 1799 and Bonaparte's failed imperial ambitions in the east. Colored etching by J. Gillray, 1799. Courtesy Wellcome Collection CC BY.

chemist, did not know how to combine volatile political ingredients in a way that prevented an explosion.

The political establishment employed chemical metaphors as well. Despite being depicted as being part of an explosion by Gillray, Edmund Burke famously attacked the comparison of legislation to experimentation. Explosions and volatile substances, averred Burke, were not positive agents of change; they were, in fact, dangerous. They were even immoral, because they inflicted great damage to the economy and stability of a country (Golinski 1992: 179). Whereas Burke attacked chemical metaphors, other conservatives accepted their use when they were couched within a providential framework in which the properties of matter and laws of nature served as examples of divine order or causation.

Put more clearly, chemical knowledge served the purpose of illustrating the overarching natural regularities issued by a divine regulator. This view was aided throughout the century by works such as Sir Isaac Newton's *Opticks*, particularly its metaphysical speculations in Query 31, which strongly suggested that a divine agent had superimposed all forces, including gravity and affinity, onto inert matter. The divinely appointed natural order, which included the substances of chemistry, served as a metaphor for civil order. The outcome was what Basil Willey once called "cosmic Toryism," an ideology that used matter theory to justify the political order (Willey 1940: 43–56).

For many experimentalists, the analogy between the movement of the planets and the movement of chemical particles, that is to say, between gravity and chemical affinity, was weak, mainly because the laws of physics did not seem to apply to the reactions they witnessed in laboratory experiments. For those more inclined to see evidence of divine or political order in the everyday combinations of matter within the human body and across the globe there was chemico-theology. This variant of the natural theology tradition advanced a view that local forces such as affinity, electricity, and magnetism were inherently calibrated to facilitate the chemical processes that sustained life. The order of such processes was taken as evidence of a divine orderer (Knight 2013). Additionally, like design arguments based on physics and astronomy, the publications of chemists such as Joseph Priestley, particularly in his *Disquisitions on Matter and Spirit* (1777), intimated that the causes that moved chemical particles were somehow influenced by a "first cause," that is to say, a divine agent. Whether referring to a divine designer or a first cause, chemical-theology provided metaphors that could be used to frame moral and philosophical discussions about the politics of individuals and institutions (Laudan 1987: 66–9).

The turmoil caused by the French Revolution during the 1790s created an atmosphere in which some chemists were forced to consider the relationship between their scientific and political beliefs. In France, Lavoisier faced a public trial and then the guillotine on account of his investment in a private tax-collecting firm for the *ancien régime*. In England, Priestley, an outspoken supporter of French republicanism, was lampooned by the press and, as mentioned above, witnessed the destruction of his Birmingham laboratory by an angry mob that disagreed with his republicanism. In Scotland, John Robison's career as an experimentalist did not allay his xenophobic fears of the atheistic undertones that he discerned in Lavoisier's matter theories (Morrell 1971). Nonetheless, there were positive political developments for chemistry as well. France's new republican government treated it as the greatest of all sciences on account of its relevance to the economy and education. French politicians gave it a prominent place in the newly designed national curriculum of "normal" (teacher training) schools, and they asked leading chemists such as Louis-Bernard Guyton de Morveau and Antoine de Fourcroy to teach French citizens how to make chemical substances such as saltpeter and hydrogen in their own homes (Bensaude-Vincent 2018).

THE ENVIRONMENT

Chemistry played a significant role in conceptualizing the environment as a system of substances, compounds, and reactions that occurred locally and globally across the terraqueous globe. Although the terms climatology, ecology, or even environmental science did not exist at the time, there was a strong sense that the earth, sea, and sky were intimately connected through material processes that had a profound effect upon humans, and upon flora and fauna more generally. The study of such processes was not new. Indeed, it had existed since ancient times. The classical physician Hippocrates, for instance, had pointed out the direct relevance of terrestrial, aerial, and aqueous cycles to human health. He argued that the hot, cold, damp, and dry fluids played a significant causal role in health and disease.

Interest in the material basis of environmental processes consistently increased during the early modern period via more specialized experimental research on rocks, soils, mountains, rivers, seas, lakes, spas, winds, precipitation, and the seasons (Glacken 1976; Porter 1980; Bowler 2000). When chemistry started to become a central part of European medical curricula during the late seventeenth century, natural history – especially the emerging fields of geology, hydrology, and meteorology – gained new importance. When these fields were introduced as courses in their own right at leading medical schools during the eighteenth century, they were called Hippocratian lectures. They provided a way to theorize the effects of the environment upon the human body (Risse 1992; Knoeff 2007; Eddy 2008).

The relationship between medicine, chemistry, and natural history significantly affected how many chemists modeled the transformation of the globe as a material system of interactive substances. Again, this interaction was facilitated by longstanding beliefs, many of which were ancient or medieval, that substances flowed through the natural world like they flowed through the human body. Prior to the eighteenth century, Aristotle's many followers believed that the elements flowed around each other to find their natural resting places in the ground and sky. The result was a conceptualization of material movement that explained everything from the growth of precious metals flowing through the strata of earth to the movement of winds across continents (Emerton 1984; Fors 2015).

From the standpoint of eighteenth-century chemists thinking about the environment, William Harvey's establishment of the circulation of the blood in the early seventeenth century provided a helpful model that allowed them to think about the movement of substances through the globe. From Harvey forward, animal and human bodies were treated as systems that circulated and transformed matter through chemical processes such as respiration and digestion. This led many to treat the globe like it was an organism. Like blood, urine, and other fluids circulating or moving within the body, the chemical composition of fluids or vapors provided insight into what was circulating inside or across the earth (Hobbs 1981).

Models of material circulation provided the conceptual foundation of theories developed by chemists who wanted to offer a natural history of the earth's formation. Such theories were underpinned by the early modern notion of an "effluvium," a liquidy substance or mixture in which material particles flowed in a manner "too subtle to be perceived by touch or sight" (quoted from Definition 2 of the *Oxord English Dictionary*). The traditional understanding of an effluvium included many kinds of watery mixtures, as well as different varieties of lava, mud, or cement. The term was sometimes applied to molten mixtures like heated glass or, in the words of the savant and Kentish priest Reverend John Lyon, to the "electric effluvia" (Lyon 1781). Many chemists knowledgeable in mineralogy, hydrology, and geology held that effluvia caused substantial changes to the structure of the globe when they were heated or cooled, or when they moved across its surface over time (Davies 1969).

Perhaps the most famous account of petrogenesis today is James Hutton's *Theory of the Earth, with Proofs and Illustrations* (1795), which was influenced by the medical dissertation that he wrote on circulation in 1749. In his reading of nature, subterranean rivers of lava combusted and circulated molten materials through the interior of the globe, the evidence for which erupted to the surface through volcanoes or appeared in the chemical composition of strata (Figure 5.4). For Hutton, the interior of the earth was like a giant engine, roasting huge masses of rock strata in a manner akin to the way minerals were concreted or fused through fire by chemists such as Johann Pott or through the scorching heat streaming through the ingenious blast furnaces invented by his friends or correspondents in Glasgow and Birmingham (Donovon and Prentiss 1980;

FIGURE 5.4 Mount Vesuvius emitting a column of smoke after its eruption on August 8, 1779. Colored etching by Pietro Fabris, 1779. Courtesy Wellcome Collection CC BY.

Dean 1992). Both inside and outside the chemical community, Hutton's theory conjured images of heat and lava in a manner reminiscent of the powerful forces wielded by Pluto and Vulcan, the Greek and Roman gods of fire, respectively, leading historians to refer to his followers as "Plutonists" or "Vulcanists."

In many respects Hutton's effluvial thesis was a newcomer to the fields of hydrology and geology. Chemists who knew how to identify the substances found in minerals promoted diluvialism, the theory that the earth was formed through waves of effluvia. Historians sometimes call theorists from this tradition "Neptunists," a name derived from Neptune, the Roman god of the sea. During this time the most prevalent form of chemical experimentation was humid analysis, that is, reactions that involved water and acids. Indeed, many of Pott's influential experiments on earths were based on these kinds of reactions (Holmes 1989: 33–60). Diluvialism offered a chemical cosmogony based directly on the results of humid experiments. According to its many adherents, the bottommost layers of the earth, present since the beginning of the world, had formed from a chemical soup that crystallized and formed a hard layer on which other floods of different mixtures concreted into strata of softer rocks such as limestone (Davies 1969).

Key diluvial books during the early part of the century included Nicolas Steno's *Dissertation on a Solid Body Naturally Contained within a Solid* (1669), which discussed the concretion and crystallization of strata, and John Keill's *Examination of Dr. Burnet's Theory of the Earth* (1698), which used biblical exegesis to offer a chemical interpretation of key words found in the creation narratives of the Hebrew and Greek scriptures. As the century progressed, diluvial theories were promoted by some of the most eminent chemists of the day, including Torbern Bergman and Johan Gottschalk Wallerius. The most influential book was *Treatise on the External Characters of Fossils* (1774), which was written by Abraham Gottlob Werner, a professor at the Freiberg Mining Academy. Diluvianism was also available to the reading public via texts written by scientific popularizers such as the Swiss-born author and traveler Jean André de Luc (De Luc 1778–1780; Rudwick 2005).

ECOLOGY

During the early modern period it was widely believed that the forces and particles of matter were somehow interconnected in a way that was guided by a divine order. The theologies of creation based on this belief fostered a divine ecology that encouraged chemists to see substances as building blocks of a supernatural designer, who superimposed forces upon matter in a way that made the earth habitable for all forms of life. Within this ecological orientation, natural disasters such as hurricanes, earthquakes, floods, and volcanic eruptions were designed to rebalance order and to inspire awe and piety (Barnett 2015).

As shown in Chapter 1 of this volume, experimental chemists often found it difficult to map their questions about the ultimate principles and forces of micromatter onto the available evidence. Add to this the fact that the relationship between personal beliefs and ecological knowledge is difficult for historians to research, mainly because the requisite evidence of personal reflections of chemists on the topic are no longer extant. That said, the notion of divine ecology was an important contextual element of early modern chemistry, especially when it comes to understanding the metaphysical and epistemological commitments that allowed naturalists to bring together environmental ideas that today would seem unrelated (Stauffer 1960; Lepenies 1982).

Experimentalists often harbored personal beliefs about matter that affected how they understood the material or metaphysical forces of the natural world (Knight 2004: 8–19). Such beliefs affected their understanding of ecology. Joseph Black, for example, marveled at the life-giving role played by the newly discovered vital air (oxygen) in his University of Edinburgh *Lectures on Chemistry* (1766). He even went so far as to say the air's properties were evidence of a final cause. Priestley's previously mentioned *Disquisitions* (1777) offered a brilliantly creative theistic account of the resurrection, in which God chemically reconstituted the bodies of all humans who had died. There was an orderly economy of matter in his ecology, and in the fact that all things could be reduced to substances that were recycled in a way that sustained life or, indeed, reconstituted the dead (Brooke and Cantor 2000: 326–9).

Through the course of the century, divine ecologies were reworked into human ecologies. Theistic materialism gave way to naturalism and increasing emphasis was placed upon the role that humans could play alongside, or independently of, a deity in designing natural or built environments. Chemical knowledge that was gained within a controlled space like a laboratory or a greenhouse served as the basis for understanding the behavior of matter on a grand scale within natural spaces like the atmosphere, rivers, forests, and more.

Water pollution is a notable example of the early relationship between chemistry and human ecology. The eighteenth century witnessed an explosion of urban living, particularly in metropolitan centers such as London, Paris, and the Low Countries. One of the most easily witnessed impacts of humans upon the environment was a decline in potable water. Here chemistry offered a number of explanations and solutions. Plant juices such as syrup of violets turned red with acids and green with alkalis, providing a way to assess the potentially harmful effects of toxic materials. Mineral well water analysis, a popular pastime of physicians living in spa towns, was applied to the rivers flowing through urban or industrialized landscapes. This led a number of chemists, Thomas Percival of Manchester for example, to argue that salts or harmful minerals were polluting the drinking water (Percival 1769; Eddy 2010).

The decline of potable water was an especially pressing problem in London, Europe's largest city at the time. Numerous physicians, including the famed William Heberden, argued that putrefied and fermented matter made the water in the New River and the Thames toxic. Several measures were taken, in London and more generally in other places, to ameliorate the problem. Meandering riverbeds, for instance, were straightened so that they could flow more freely, which was believed to purify the water. Swampy areas were drained. Sewage from factories and manures from farms were monitored. Vegetation, particularly leafy trees, was cleared so that it could not fall into the water and rot (Tomory 2017).

The chemistry of human ecology played an important role in understanding the causes of soil depletion as well. Ever since antiquity it was widely known that soils ceased to be fertile if they were overused. Infertility increasingly became a European problem in the eighteenth century as the population increased, or as imperial policies encouraged the overuse of land in colonies. Farmers increasingly turned to chemists to help solve the problem. Efforts were made to prevent soil depletion by developing new fertilizers based on local materials or encouraging crop rotation, including periodically leaving fields fallow. In Britain and the Netherlands, the effort was sometimes ad hoc, organized by chemistry professors based in universities. In other countries the effort was coordinated more by the state. The Berlin-based pedagogue and pharmacist Sigismund Friedrich Hermbstädt, for example, worked diligently to integrate *Kammeral-Chemie*, state-sponsored chemistry, into the agricultural practices of Prussia (Hermbstädt 1808; Eddy 2007; Egerton 2012: 175–7).

With the coming of the Industrial Revolution chemical reasoning was applied to questions relating to urban air quality and its connection to meteorological conditions. Smoke in particular contained a plethora of caustic substances such as fumes, soot, and smog, the effects of which were exacerbated by the domestic and public use of fashionable tobacco products (Figure 5.5). Coal and tobacco smoke destroyed the drapery, furniture, paper, and leather found inside homes, hospitals, ships, workhouses, and prisons, making it possible for denizens to literally watch its adverse effects materialize before their eyes. If this was happening to the interior of buildings, what was it doing inside the human body? Combine this widespread suspicion with the fact that there were influential medical theories that attributed the spread of disease to bad airs, "miasmas," that made stagnant air lethal. Such theories fell within the domain of medical chemistry and encouraged entrepreneurs to invent smoke-resistant upholstery, savants to design ventilators, and physicians to propose architecture that purified the atmosphere of a building through the movement of air (Hales 1743; Adair 1790: 26–59; Brimblecombe 2011: 63–89).

FIGURE 5.5 Four men sit round the tax man and blow smoke in his face. Colored aquatint, late eighteenth century, after G. M. Woodward (?). Courtesy Wellcome Collection CC BY.

The chemical ecologies that emerged in the latter half of the century were significantly influenced by the discovery of different kinds of gases. Although air had been seen as an aerial fluid for quite some time, it was generally treated as a homogeneous substance that carried heat and vapor across the globe in a manner analogous to water. Consequently, the main goal of early Enlightenment meteorology was to measure atmospheric temperature, pressure, and precipitation. This program expanded after the isolation of fixed air (carbon dioxide), vital air (oxygen), and inflammable air (hydrogen) from the 1750s onward. Experiments with plants and animals inside closed containers revealed that vital air sustained life, while fixed air and inflammable air ended it. In addition to identifying a cause that prevented respiration in enclosed areas like mines or diving bells, pneumatic chemistry helped savants, architects, and literate home owners alike to reconceive the circulation of foul airs or miasmas (Jankovic 2010).

The relationship between plants and gases in particular triggered a reconceptualization of the material cycles that underpinned ecological systems. More specifically, in a series of highly popularized experiments, Joseph Priestley revealed that plants possess the capacity to somehow transform fixed air, which did not support respiration, into vital air, an essential gas that sustained all known life. Conversely, he revealed that humans and other animals transformed vital air into fixed air when they breathed. In short, not only were gases somehow recyclable, there was a direct form of material interface between the organic and inorganic world. Other eminent chemists, Antoine Lavoisier, for example, recognized the ramifications of Priestley's work and conducted experiments that sought to further understand the ecology of airs and their relationship to health, climate, and the environment (Priestley 1790; Beretta 2012c).

CONCLUSION

Founded in 1780, the stimulating chemical discussions of the Baptist's Head Coffee House Philosophical Society in London came to an end already in 1787. Its members were still interested in chemistry, but the knowledge they had gained in the meetings opened new sociable and commercial opportunities. As we have seen in this chapter, this kind of adaptability and utility was a hallmark of chemistry during the long eighteenth century. It allowed it to permeate many levels of society, influencing everything from the development and demand of products to the political commentaries issued by print and visual culture. It also enabled some chemists to support themselves through a variety of middle-class occupations. At the same time it led aristocratic, governmental, or industrial patrons to fund large-scale research and development projects. Finally, and perhaps most noteworthy for its historical legacy, chemistry's versatility impacted the ways in which the environment was conceptualized as a dynamic entity, concurrently leading to destructive and ameliorative ecological ideas advocated by individuals and institutions.

CHAPTER SIX

Trade and Industry: *An Era of New Chemical Industries and Technologies*

LESLIE TOMORY

INTRODUCTION

The eighteenth century was transformative in Europe for what would later be considered chemical industries. Indeed, one important new factor present among observers of trades was a greater awareness that chemical trades included a broader range of activities than had previously been classified as "chemical" or "chymical." Although never to the exclusion of others, the trades that were considered chemical were usually in metallurgy or more especially in the apothecary trades, with alchemy disappearing from the scene early in the century. By 1800, numerous authors, such as Gottfried Hoffmann in his 1774 *Anleitung zur Chemie für Künstler und Fabrikanten*, had demoted pharmacy to one among many chemical trades.

This emergence of chemical industries in chemical literature after 1750 paralleled the changes taking place in contemporary industrial activity. In the early eighteenth century the vast majority of trades that would later be considered chemical industries, such as textile bleaching, were traditional. They were carried out by artisans learning skills on the job through apprenticeships, typically in small workshops, laboratories, or other sites of production. They were run by individuals, using relatively little fixed capital, and the growth of knowledge and sources of innovation were largely internal to the trade. Although there was innovation in business practices and technology during the century, this often originated within the trade itself. Internal changes of this nature certainly continued, but as the eighteenth century progressed, chemical trades were also affected by external factors such as the growth of international trade, changes in patterns of consumption as more people desired a greater variety of goods, and new possibilities for chemical processes emerging from speculative chemistry and the interventions of mercantilist states, such as in France, Prussia, and Sweden. The desire to imitate oriental goods, for example, prompted important changes in European textile and pottery production, such as the foundations of large porcelain manufactories. A host of new materials and knowledge of their properties came from academic chemistry, especially after 1750. Although much of this knowledge would affect industries only after 1800, its influence was already visible earlier in industries such as chlorine bleach and alkali production.

TEXTILE PRODUCTION

Manufacturing textiles includes many chemical processes, notably the preparation of fibers, followed by their bleaching and/or dyeing, with the concomitant fixation of the dyes to the fibers. All these steps vary with the fibers being used, the final type of fabric being produced, and the type of dye being applied. These steps were done in many ways in the eighteenth century, which was a dynamic period for the textile industry. The mechanical transformation of the industry has long held a prominent place in the history of the Industrial Revolution, with inventions such as the spinning jenny (1760) and power loom (1785) among others helping to make cotton the leading fabric in terms of volume, and securing Britain's pre-eminence in the industry. However, the changes in the chemical aspects of the textile trade were also far-reaching. In the early part of the century, most of the industry was traditional. Almost all the dyes, bleaches, and other chemicals present at the beginning of the century were still in use at its end.

However, new chemicals were introduced to the textile trades, and new ways of producing and applying chemicals also emerged. These transformations were prompted by innovations internal to the industry itself, but they also came about as a result of shifts in the broader economy, such as the mechanization of production and the growth in colonial trade. Mechanization increased the volumes of textiles produced, boosting demand for the chemicals used by the industry, which in some cases strained the traditional sources of supply. Colonial trade also introduced new materials to the industry, such as dyes, as well as opening new sources of raw materials and markets, for example cotton from the Americas. The ways materials were sold also changed, as the production of chemicals shifted from small shops to heavier manufacturers (Nieto-Galan 2001).

BLEACHING

In the early eighteenth century, the process of bleaching included a number of stages, some of which used chemicals. The process began with the boiling of the fabric in water containing potash or pearl ash (mostly potassium carbonate) to remove fats and oils; washing it with soap; and finally souring it with a mild acid. Potash was typically made from vegetable ashes, which were also used for many other processes (see the discussion of alkalis below). The souring liquors included sour milk and buttermilk. Souring removed salts from the fabrics, allowing them to hold their color better once dyed. These steps had to be repeated a number of times. Once treated this way, the fabrics were left in a field exposed to the sun for several months before they attained the necessary degree of whiteness to receive dyes or to be marketed as white fabrics. The process required lots of water as well as field space for fabrics to lie in the open. Until the mid-century, Holland was a traditional center for bleaching of linen for the luxury market, and it imported many raw fabrics to be bleached before re-exportation. Ireland and Scotland also developed bleaching industries in the eighteenth century (Clow and Clow 1952: 178–80; Nieto-Galan 2001: 65–6).

Before the 1750s, changes in chemical inputs to bleaching came about as new sources of potash were developed, notably the seaweed kelp and barilla (see below). Another innovation introduced in Scotland around the 1720s was mixing lime (calcium hydroxide) with a potash solution to produce caustic potash (potassium hydroxide). This was a stronger detergent than potash alone. The caustic potash solution, however, was powerful and harmed the quality of the cloth. To protect the industry's reputation against charges of poor-quality textiles, several laws were passed banning it soon after it was introduced, although it was back in use by the 1760s (Clow and Clow 1952: 183–4; Gittins 1979).

In the latter half of the eighteenth century the bleaching process itself began to change as two new chemicals were introduced: sulfuric acid and chlorine. Sulfuric acid was a relative newcomer to widespread commercial use in the eighteenth century. Other sulfur compounds had been used in earlier periods as mordants for dyes. Copperas, or green vitriol (iron sulfate), was used in this way from the sixteenth century. "Oil of vitriol" (concentrated sulfuric acid) was used in pharmaceutical applications, but it was difficult and expensive to produce in large quantities. There were two ways used to make concentrated sulfuric acid before 1750. The first involved distilling green copperas and condensing the rising acid vapors. The second was to condense acid fumes with a glass bell placed over burning sulfur, producing "oil of sulfur." Apothecaries had known since the seventeenth century that a little saltpeter (potassium nitrate) could increase the yield of the second process, but this was not generally done because it was thought to introduce impurities (Smith 1979: 5).

The regular production of sulfuric acid began when the physician Joshua Ward and his assistant John White set up a works at Twickenham in Britain around 1736. They used large glass globes in which they burned sulfur and saltpeter. The fumes condensed in the globes and ran off into water at the base of the apparatus. The price of sulfuric acid from their workshop dropped to around a tenth of what it had been. Ward's plant seems to have stopped operating by 1749, but a further innovation brought much larger volumes. In 1746, John Roebuck and Samuel Garbett began producing sulfuric acid at their plant in Birmingham using a new process in which acidic fumes were condensed on the walls of a lead chamber. They soon established a larger plant in 1749 in Prestonpans, near Edinburgh. They were able to produce sulfuric acid in much larger quantities and the price of acid dropped again. They did not patent the process, and within a few years competitors emerged using the lead chamber process. Sulfuric acid was initially produced for apothecaries and metal refiners. Its use was extended as a souring agent in the bleaching of textiles from 1753 at the suggestion of Francis Home, an Edinburgh physician. It was soon adopted by bleachers in Britain. The use of sulfuric acid changed the bleaching process so that the time required was reduced the from six to eight months per batch to a day or so, and therefore many acres of land were no longer needed to bleach fabric (Fester 1923 [1969]: 142–3; Clow and Clow 1952: 181–3; Gittins 1979: 195; Smith 1979: 6).

Sulfuric acid was also produced in France, when the Englishman John Holker set up works with government help in Rouen in 1769, relying first on glass globes, and then from 1772 lead chambers. His son had apparently learned of the lead chamber process on a visit to Scotland. In the first year of production, Holker sold 45,000 lbs (20.4 tons) of acid and by the 1790s he was selling 300,000 lbs (136.1 tons) per annum. Two improvements were made to the lead chamber process in France. The first was proposed in 1774 by the apothecary Louis-Guillaume de la Follie and consisted of spraying water into the lead chamber. The second, perhaps suggested by the chemist Jean-Antoine Chaptal in the 1780s, was to force air into the lead chamber to make it a continuous production process (Smith 1979: 7–12, 15, 69, 77).

Another new chemical introduced into the bleaching process in the eighteenth century was chlorine as a bleaching agent. It came into use soon after the Swedish chemist Carl Scheele's discovery of chlorine in 1774. In a 1785 paper, Claude-Louis Berthollet noted its possible application for bleaching. It was not easily adapted for commercial application, however. Preparing, storing, and applying chlorine gas was difficult and dangerous. It would corrode metal containers and was poisonous. Despite this, it was adopted and by 1789, a number of bleachers in Scotland and Lancashire were using it. Chlorine was also used in France after 1789, but the most significant development there was a liquid bleach called "eau de Javel". It was first made by a sulfuric acid manufacturer established in 1778 in the town of Javel, outside Paris. The Javel works had been making hydrochloric acid from 1780, and when Berthollet made known his discovery of the action of chlorine bleach in 1785, the works owners saw a new market. They began to market the liquid bleach, a stable, shippable commodity which was much easier for bleachers to use because it spared them having to produce chlorine gas themselves. They prepared the bleach by passing chlorine through a potash solution, producing a solution of what today is known as potassium hypochlorite. Unfortunately for the company, bleachers soon preferred to produce the solution themselves rather than purchase it (Smith 1979: 18–19, 129–32; Nieto-Galan 2001: 67–9).

Chlorine bleaching was made even easier when the Scottish bleacher Charles Tennant discovered how to produce a powdered form of the chemical. He took out his first patent for producing a liquid chlorine bleach in 1798, and followed this with a second one for a powder in 1799 based on his partner Charles Macintosh's earlier observation that wet lime absorbs chlorine to form "chloride of lime" (calcium hypochlorite). When dissolved in water, the powder (like "eau de Javel") releases chlorine for the bleaching effect. The powder was much lighter than the liquid solution and hence cheaper to ship. The firm of Charles Tennant and Co. established a plant in St. Rollox in Scotland that same year, and by 1830 it was the largest chemical works in Europe (Clow and Clow 1952: 190–3; Christie 2017). However, the importance of the powder and the works lay in the nineteenth century.

DYES

In the early eighteenth century dyeing was an artisanal pursuit based on natural dyes. The work was done in workshops of ten to twenty people overseen by a master dyer who was a member of a dyers' guild. The principal dyes included red cochineal derived from a cactus-feeding insect found in the Americas and elsewhere. The blue dye indigo was produced from plants native to India but later introduced to the Americas and beyond. A similar blue was also made from woad, a plant grown in Europe. Other vegetable dyes were madder, logwood, campeachy wood, brazilwood or redwood, yellow quercitron from oak, and the yellow dye weld. All of these were used industrially throughout the eighteenth century and some even as late as 1910. Interest in new dyes and techniques to apply them was strong. In France, the Gobelins manufactory was a key site for experimentation (Figure 6.1). Established in the 1660s to produce tapestries, it also developed new ways of applying dyes such as weld, madder, cochineal, and indigo. From the mid-eighteenth century, it housed a school and laboratory.

FIGURE 6.1 Gobelins Dye Works, from *Encyclopédie, ou dictionnaire raisonné des sciences, des arts et des métiers*, eds Denis Diderot and Jean le Rond d'Alembert (1772), vol. 10, plate VIII, colored tint from ca. 1800. Sourced from Wikimedia Commons, Public Domain.

Many academic chemists, including Charles François de Cisternay du Fay, Jean Hellot, Pierre-Joseph Macquer, and Claude-Louis Berthollet worked there and built up its international prestige (Fairlie 1965; Nieto-Galan 2001; Lowengard 2008).

Many dyes had grown in importance in international commerce before the eighteenth century. Up to the seventeenth century, Europeans observed that India produced spectacularly colored textiles not found at home, especially cotton. Cotton cloth in India was decorated by printed and painted designs, rather than made by dyeing and weaving, which was the standard method for wool, the dominant fiber in Europe. The ability of European textile manufacturers to produce cotton textiles gradually improved so that by the 1780s European callicoes and other cottons were superior to Indian textiles. This shift in status was the result of many changes, especially the mass production of cheap cotton cloth, but also included the creation of new ways of dying textiles. Indeed, the shift to cotton was not only about new fabrics and colors; cotton production became a locus for the factory system and mechanization of fabric production.

Mechanization also extended to the printing of calicoes in the 1780s following Thomas Bell's design of roller printers (Nieto-Galan 2001: 52, 56–7; Riello 2010).

Indian techniques for printing indigo on cotton were likely imported to Europe by Armenian dyers in the seventeenth century. European dyers found new methods of applying the dye, the most important of which involved dissolving indigo with iron sulfate placed in cool vats. The process was developed in England in 1734 and did less damage to the textiles than previous techniques. Another new process was to print indigo using potash, quicklime, and orpiment (arsenic sulfide), introduced in the 1730s and 1740s. "Turkey red," derived from the roots of the madder plant, was another dye imported to Europe in the seventeenth century. At first, dyeing with madder in Europe was only done poorly so that even around 1750, Turkey red-dyed cloth was mostly still imported. In the 1740s and 1750s, French and English merchants recruited dyers from the Ottoman Empire and established the trade in Europe, although for decades thereafter Europeans continued to import the dye and to seek to better understand its nature and properties. European varieties of the madder plant were increasingly cultivated from the 1790s (Nieto-Galan 2001: 20–1, 186; Riello 2010: 14–17, 20–3).

Turkey red was one of many eastern dyes that European chemists sought to analyze and reproduce, as well as to improve their preparation and use. The first natural dye to be modified chemically came with the accidental discovery around 1706 of Prussian blue (iron ferrocyanide) by an obscure Berlin dyer, Johann Jacob Diesbach. Partly by happenstance, he discovered this deep blue dye through a complex process starting with bull's blood and pulverized horn. Its recipe was kept secret until 1724, when John Woodward published an account of its preparation in the *Philosophical Transactions*, and its production soon spread throughout Europe. It was mostly used as a paint pigment, and later as a dye for silk and calico printing. Another dye, picric acid, was created in 1777 by the Irish chemist Peter Woulfe when he decomposed indigo with nitric acid. It was not much used until the nineteenth century (Nieto-Galan 2001; Kraft 2008; Lehman 2012: 309).

MORDANTS

Mordants are usually metal salts, such as copperas and cream of tartar (potassium hydrogen tartrate), but by far the most commonly used mordant was alum (a group of related aluminum sulfates), which was produced from shales or slates by roasting for several months and then washing to extract the salts (Figure 6.2). The resulting solution was then boiled with certain other substances, like seaweed and urine, and the product crystallized out. From the fifteenth century, the production of alum was a papal monopoly thanks to the mines at Tolfa, near

FIGURE 6.2 Alum works, from *Encyclopédie, ou dictionnaire raisonné des sciences, des arts et des métiers*, eds Denis Diderot and Jean le Rond d'Alembert (1768), plate, vol. 5, p. 26. Sourced from Wikimedia Commons, Public Domain.

Rome. However, this monopoly was broken when the slate roasting process was developed in the seventeenth century by the Protestant countries, Sweden and England. Alum came to be manufactured on a large scale in Scotland when Charles Macintosh began producing it from shale tailings of coal mines in the 1790s. The tailings had built up in the area around Hurlet, near Glasgow, over the course of 200 years. Earlier experiments made in the 1760s and 1780s to transform the tailings into alum had been unsuccessful, but in the 1790s Macintosh succeeded in creating a viable process. It consisted of leeching metal salts out of the shale, which had lain exposed to the air, oxidizing and acidifying its sulfur over the years. The salt solution was then concentrated into copperas and alum after the addition of potash. Heating and cooling the alum further concentrated it to a level sufficient for sale and transportation (Clow and Clow 1952: 235–40; Fairlie 1965: 492).

Like alum, copperas or green vitriol had been used since antiquity mostly as a mordant, but also for making inks and acids, including sulfuric and nitric acids. Its use in Europe grew after the sixteenth century and proliferated in the seventeenth century. The principal production process involved oxidation of iron pyrites (iron disulfide). Pyrite stones were collected, often from alluvial deposits, and left to weather for several years, during which time they produced a liquor (a mixture of iron sulfate and sulfuric acid) that was collected in barrels. This liquor was mixed with scrap iron, which reacted with the sulfuric acid and was boiled for many days, in a process that used huge quantities of fuel. The copperas was then left to cool and solidify. In England, small copperas works employing only a few workers were located at Deptford and Blackwall on the Thames in the early eighteenth century. Production shifted towards the north from 1750 to Yorkshire, Tyneside, and Scotland (Allen 2001: 96–101). Production of copperas in England slowly grew, until around 1700 when it was exported for the first time. By 1764, England was the largest European producer, exporting over 2,000 tons per year by the 1780s.

METALLURGY AND MINING

The metal industries were shifting geographically throughout the eighteenth century. In the seventeenth century, central Europe was the predominant area of production and relevant skills. The devastation of the Thirty Years' War depressed the local economies, and regions outside of central Europe gained in strength, particularly Sweden and Britain. Large inflows particularly of precious metals from the Americas also produced marked shifts. As in prior centuries, iron was the most extensively used metal of the eighteenth century. The volume of production and the range of its use expanded notably. Non-ferrous metals, however, also saw important developments.

Copper

In the early modern period Germany dominated the production of copper, used not only as copper metal itself but also in brass and bronze production, and German techniques were exported to other areas in Europe. Copper was made by roasting copper sulfide ore in shaft furnaces, described in the sixteenth century by Agricola. Sweden provided the majority of Europe's copper at the beginning of the eighteenth century, mostly from a deposit near Falun. However, German and Swedish production was diminishing while English production was on the upswing, as copper was found associated with Cornish tin deposits. By 1700, around 1,000 tons of Cornish copper were being smelted in Bristol and Gloucester, an amount comparable to Sweden's output. Britain passed Sweden to become Europe's largest producer around 1720. Spain also imported large quantities of copper from the Americas. Between 1761 and 1775, 3,450 tons of copper were transported to Spain from Peru. There were two English technical innovations in copper smelting that aided the local industry's expansion: the use of coal, and the reverberatory furnace, in which the copper was not in direct contact with the burning fuel. The use of reverberatory furnaces was adapted to copper from lead smelting in 1700. It was a complex ten-stage process that first separated the iron from the copper in the ore, and then purified the copper of sulfur and other impurities. Most of the copper produced was used to make brass, as well as copper sheathing to protect the bottoms of ships (Fester 1923 [1969]: 135; Tylecote 1992: 109–10, 149–51; Burt 1995: 27).

Zinc and Brass

Before the eighteenth century, zinc was mostly used for brass as an alloy with copper and occasionally other metals. Its ore was usually found together with other metals, such as in lead ore deposits in the Harz Mountains. Brass was traditionally manufactured throughout Europe without first isolating zinc metal by heating copper, zinc ore (calamine), and charcoal in a crucible, sometimes with admixture of alum and salt. The molten brass was then poured into larger crucibles to produce ingots. This cementation process based on zinc ore could not produce brass of greater than 40 percent zinc that was desired for many applications. Such high-zinc alloys required admixture of metallic zinc, but even well into the eighteenth century Europeans had not yet mastered the technology necessary to make metallic zinc. Instead, European artisans were forced to import expensive zinc metal from India, where an advanced metals industry had long since mastered zinc smelting.

Brass was used to produce household implements such as candlesticks, and navigation and surveying instruments, but also military items, including cannon. Brass was also drawn into wire and was used to make pins that were later coated with tin. It was manufactured throughout Europe. Bristol grew to be a center of brass production in the early eighteenth century when Abraham Darby and his partners established a brass works there in 1700, using skilled Dutch immigrant workers. By 1712, they were producing around 250 tons of brass annually. Much of the brass was sent to Birmingham where it was worked into finished products and then exported (Day 1991; Tylecote 1992: 151–3; Darling 2002: 84–7, 91).

Metallic zinc was rarely used, at least in Europe, but this changed after William Champion, a brass-maker in Bristol, developed a production process using a vertical furnace in 1738, presumably inspired by similar Indian technology and by local brass and glass furnaces. He patented the process and opened a zinc works at Warmley near Bristol. The challenge in smelting zinc was that the zinc oxide from the ore vaporizes before it is reduced by carbon into zinc metal. The zinc needs to be condensed after the reduction reaction has occurred and before it comes into contact with air. Champion's furnace contained iron tubes to capture and then cool the zinc vapor. Enterprising visitors came to examine and copy his process. New, more efficient variations of his furnaces that used hundreds of smaller tubes were introduced by Bergrath Dillinger around 1799 at Dollach in Carinthia in Austria. Johann Ruberg further improved the process when he built his zinc smelting works in Wessola (now Wesoła) in Upper Silesia in 1798. His horizontal furnace could be recharged without having to be cooled. New applications were found for zinc in the course of the eighteenth century; for instance, in 1742 Paul-Jacques Malouin of Rouen discovered that coating iron with zinc made it resistant to corrosion (Moulden 1916: 498–9; Tylecote 1992: 151–2; Darling 2002: 95; Craddock 2009).

Tin and Lead

In the eighteenth century, with the decline of eastern European centers of production, Cornwall was the major source of European tin, although much was also imported from China and elsewhere in the east. Tin was produced by first grinding the cassiterite ore to separate it from siliceous gangue, and washing it. It was then smelted in blast furnaces, where the reduced metal and slag were separated. Originally, wood, charcoal, or peat was used for fuel, but in England coal was introduced in the early eighteenth century, as was the reverberatory furnace. Initially, much of the tin used in Cornwall was alluvial, but as steam engines pumped the water out of ever deeper mines, vein tin became more common. However, vein tin contained pyrite impurities and so needed preliminary roasting in a kiln. Coal-smelted vein tin was of relatively poor quality, but Cornish smelters managed some improvements as the century progressed. Tin was mostly used for making bronze by alloying with copper, but new applications also emerged, notably the tinplating of iron. Tinplated iron became a popular substitute for glass and ceramic articles. As European Continental tin production declined, British tinplating grew in importance. Tinplating had been introduced to Britain from Saxony in the seventeenth century. Tin was also exported to China (Earl 1991; Tylecote 1992: 116, 159–61; Burt 1995: 37).

The British expansion of tin and copper production in the eighteenth century was based on an earlier growth in lead production, and Britain was the largest producer of lead during the seventeenth and eighteenth centuries. By 1700, Britain was producing tens of thousands of tons of lead per year. Some of the techniques that became important in tin and copper metallurgy, such as the reverberatory furnace, were adopted from lead production. Lead was used for a wide range of products, including ammunition, but also for roofs and pipes, as well as in paint, glass, and other products, including containers for shipping and storing goods.

Lead was also used to produce the pigments white lead (lead carbonate) and red lead (lead oxide). White lead, used in plaster, pottery, and paint, was predominantly manufactured in the Netherlands from lead imported from Britain. It was made by casting the lead into thin sheets and immersing these sheets in pots filled with vinegar. Many pots were then stacked in manure, and as the manure fermented it heated the vinegar, which evaporated and corroded the lead. The white lead was separated from the uncorroded lead, washed, and dried. This "Dutch process" was also used in Britain, and spread elsewhere in Europe in the eighteenth century, although the Netherlands continued to dominate. By the end of century, the Dutch white lead industry produced 4,000 tons per year, mostly for export. The process of pulverizing the white lead was very dangerous for workers. Machines to break the white lead under water were introduced by Archer Ward in Britain after 1750, and this made the process much safer. Furthermore, in 1787 one of Ward's partners, Richard Fishwick, replaced the manure with spent bark from tanneries, and found that it made the process more manageable (Rowe 1983: chap. 2; Willies 1991; Burt 1995: 27–9, 32–4; Homburg 1996: 32–9).

Iron and Steel

The iron industry in the eighteenth century expanded dramatically with the introduction of a number of new techniques, especially in Britain. In the eighteenth century, different sorts of iron were produced with varying carbon content: pig iron (>4 percent); cast iron (2–4 percent) used to make larger items such as cannon; steel (0.5–2 percent, see below); or the softer wrought iron (<0.5 percent) for smaller items, such as nails. At the beginning of the eighteenth century smelting was done in blast furnaces burning charcoal. The process of refining involved melting and reducing the iron, while drawing off the impurities by means of a lime flux to aid the formation of slag. These furnaces produced small quantities of iron because the charcoal could not be piled high in the furnace, limiting how much iron was smelted in a single charge. Sweden was the most important producer of iron in the early part of the century (Hyde 1977: 7–17).

Abraham Darby, a brass- and iron-founder in Bristol, leased a blast furnace in Coalbrookdale in 1708, and by the 1720s he was using coke (roasted coal) to smelt iron. Coke was similar to charcoal in its properties, but was less expensive than charcoal, and did not compete with other important uses for wood. Although coke furnaces could be built much larger than charcoal ones, coke smelting did not spread until after the 1750s because pig iron from coke furnaces was high in silicon. This meant that it was more expensive to refine and rework as bar and wrought iron. Darby's production of thin-walled cast iron products was economically viable, but it was not until charcoal prices rose and those of coal fell that many more iron-founders in Britain switched. Their pig iron was soon cheap enough to be attractive to iron-forgers. Coke smelting spread to other countries only in the nineteenth century (Hyde 1977: 27–8, 38, 57–62).

Further technological change in the industry came with the use of steam engines of Matthew Boulton and James Watt's design to drive blast furnaces after 1780. More important was the introduction of coal to the refining of wrought iron. The Wood brothers (Charles and John) introduced coal refining in the 1760s with the potting process. This consisted of burning the silicon out of pig iron with coal. The iron was contaminated with sulfur from the coal in the process, but this was in turn removed by melting the iron in pots with a lime flux, which absorbed the sulfur. A further significant improvement was the puddling process, which used a reverberatory furnace to melt the iron while a puddler helped remove the impurities by stirring with an iron staff. These changes allowed the output of British iron to increase markedly to the end of the century (Hyde 1977: 70–3, 83, 88).

Steel had long been prized because of its hardness relative to wrought iron, without the brittleness of cast iron. It was, however, difficult to produce iron with the correct carbon content to make steel, not the least because the chemistry of steel was not well understood. During the eighteenth century a number of chemists sought to understand it better. In 1722, René de Réaumur observed that iron placed in charcoal absorbed salts and sulfurs. In 1781 Torbern Bergman noted that the quantity of "plumbago" (crude graphite) in iron accounted for the observed variation in properties. Steel, he suggested, comprised between 0.2 percent and 0.8 percent plumbago, an observation reinterpreted in 1786 in France by Alexandre-Théophile Vandermonde, Gaspard Monge, and Berthollet as being its carbon content.

Steel had been made since antiquity, but in Europe by the early eighteenth century, it was produced either by melting cast iron to burn out enough carbon to form natural steel, or more usually by taking wrought iron and re-carburizing it by heating it in crushed charcoal to form "blister steel." These processes were very sensitive to conditions; producing any quantity of steel took a great deal of skill and experience. Even so, the steel was uneven in its carbon content and contaminated with impurities. Blister steel bars were welded together in forges to even out the quality of the steel in final products. New techniques for steel production emerged in the course of the eighteenth century. In the 1740s, an English toolmaker in Sheffield, Benjamin Huntsman, developed the crucible process. It entailed melting blister steel in clay crucibles placed in a coke furnace. A "secret flux" was also added to collect impurities in a slag that was skimmed off. Huntsman managed to solve the problem of bubbles forming in the steel to produce cast steel that could be reworked by smiths who purchased the ingots. Initially, Huntsman exported most of his steel to France, but because he had no patent his techniques were copied by others in England and then elsewhere (Smith 1964; Barraclough 1991).

DOMESTIC GOODS

Historians have noted that domestic consumption was responsible for significant economic changes in the eighteenth century. The demand for cotton textiles such as calicoes mentioned above is an example of how preferences sustained the growth of new industries that were central to the industrialization of Europe. Although not as dramatically as cotton, the production of other household goods increased in the eighteenth century (Berg 2004; Berg 2010; De Vries 2008).

Pottery and Glass

The desire for new forms of earthenware, inspired by oriental patterns, had already prompted experimentation before the eighteenth century. Dutch potters had been making Delft ceramics using blue tin-enamel glazes on creamy white stoneware clay from the seventeenth century. In France, new soft-paste porcelain that required lower firing temperatures was manufactured from 1693 at St. Cloud and from 1745 at a factory in Vincennes, which later moved to Sèvres in 1753.

The first true porcelain was made by the apothecary Johann Friedrich Böttger in the town of Meissen around 1710, where he worked for the Elector of Saxony. After ten years of experiments, he discovered that hard-paste porcelain could be made from a local clay, kaolin, and feldspar. This ceramic material vitrified when fired, yielding a translucent surface sheen. Knowledge of the technique spread to the French royal manufactory at Sèvres. Hard-paste porcelain was made there from 1766 when Macquer, who held the post of academic chemist at the works, found a suitable clay based on alum earth. Hard-paste porcelain was also introduced to Prussia and England. It was first made by William Cooksworthy in Plymouth in 1768, but it only flourished when Josiah Wedgwood began its production in 1785. Wedgwood, based in England's pottery district of Staffordshire, had been an innovative ceramicist for some time. Inspired by the English potter John Astbury's introduction of lead glazing before 1740, Wedgwood had in the 1760s been developing lead-glazed, cream-colored earthenware known as creamware, as well as improving colored glazes. In 1777 he invented jasper, a white stoneware that could be stained by metallic oxides. It was used to make two-color bas-relief cameos. In 1782 he also designed the pyrometer, a device to ascertain the heat of kilns. His introduction of machines to expand production was as important as his chemical contributions (Day 2002: 192–3; Dolan 2004; Lehman 2012: 317–32; Klein 2013; Klein 2014a).

An important site of manufacture of glass in France was a large manufactory founded under Colbert in 1665, and transferred to Saint-Gobain in 1693. In addition to production of household glassware by blowing, large glass panels were cast there and used for mirrors and windows. René de Réaumur experimented with glass and in 1740 managed to imitate Chinese porcelain with white opaque glass. The discovery was to spark further French research into glass at the Sèvre manufactory, especially as glass was increasingly used in scientific research in astronomy, chemistry, and other fields. Another new form of glass was the lead oxide or flint glass. It had been invented by the English glassmaker George Ravenscroft in 1674. This glass was easier to melt and shape, maintained a sparkling transparent appearance, and could be melted over a coal fire in pots. The technique spread to other English glassmakers and then to European glassmakers in the eighteenth century. Red lead (lead oxide) was frequently the source of lead, increasing the demand for this substance. A major constraint on glass production remained the availability of alkalis, typically made from kelp and barilla. This was not resolved until the introduction of the Leblanc process (see below) (Clow and Clow 1952: chap. 14; Beretta 2012b).

Soap, Candles, and Oil

Soap and tallow candles were among the most commonplace of consumable household goods. Manufacturers were ubiquitous, and the processes that produced them varied according to the local practice and available source materials. Soaps are produced by boiling fats with alkali, with glycerin as a by-product. Many different sources of fats were used, including animal fats and vegetable oils, as well as different kinds of alkali, such as plant ash. Soft soaps were made with potash, while soda produced harder soap. Additives could also change the soap's qualities, such as pine resin giving yellow soap. The addition of salt during production to speed the process of separating out the soap from the solution was introduced before 1700. In the eighteenth century, European chemists were interested in improving the quality of soap by the addition of new substances. For example, Nicolas Lémery suggested that fillers could be added to soaps to make them whiter and give them more body. By the end of the century, soap makers were adding materials such as talc, rosin, silica, starch, sodium sulfate, common salt, and borax. Soap was also linked to a global trade in materials, with purer animal fats being exported from Russia and the Americas, and palm and olive oil from warmer regions, including Africa and India (Gibbs 1939: 177–80; Schranz 2014: 258–9).

Most artificial light during the eighteenth was produced either from candles, most of which were made from inexpensive tallow, or from oil lamps. Tallow candles were largely made at home by hardening animal fat around wicks. Beeswax gave a clean-burning flame, but was very expensive and only used in limited circumstances, such as church candles. In the early eighteenth century, oil from sperm whales began to be used in larger quantities in lamps. It had been known for centuries that fish and whale oil could be used as an illuminant; the whale-oil trade expanded sharply in the eighteenth century with improved ships. Sometime before 1750, it was discovered that a waxy substance from sperm whales, spermaceti, made excellent candles. The use of whale-oil lamps and spermaceti candles then expanded rapidly after 1750, for example, in lighthouses. Unlike tallow, making spermaceti candles was not a domestic process. Wax had to be separated from oil mixed with it in repeated steps and then boiled before being hardened. This was done in factories (Dolin 2007: 46, 54, 101, 120; Irwin 2012: 52).

The end of the eighteenth century also witnessed other improvements to lighting. Aimé Argand invented his eponymous lamp with its more efficient cylindrical wick in 1780. However, more important in the long run was the development of lighting with coal gas. This technology was invented simultaneously by a number of people, notably Philippe Lebon in France and William Murdoch in England in the 1790s. It consisted of gasifying coal in a low-oxygen oven and piping the evolved gases after purification to customers' lamps. The technology would grow into city-wide networks only after 1800.

Phosphorus matches were another invention of the eighteenth century that would flourish especially after 1800. Phosphorus had been discovered in 1669 by the alchemist Hennig Brand in Hamburg. Because phosphorus ignites easily, it was tried as a substitute for flint in starting flames, but it proved to be difficult to make in quantity. It was easily available only after bone ash (calcium phosphate) came into use for the production of bone china in the late eighteenth century. Carl Scheele had observed in 1775 that phosphorus could be made from bone ash with sulfuric acid. The first match-like devices using phosphorus were the French phosphoric taper (1781) and the Italian phosphorus bottle (1786). These contained phosphorus-tipped tapers that ignited when rapidly withdrawn from containers. They were largely curiosities, but reflected ongoing experimentation that would produce useable matches in 1803 (Clow and Clow 1952: 449–52; Schrøder 1969; Wisniak 2005; Tomory 2009).

Agriculture and Food

The most domestic of industries is the growing and preparation of food, which involves much chemistry, from fertilizers for agricultural production, to manipulating biochemical processes such as fermentation, to actual cooking. Synthetic fertilizers would not arrive until the nineteenth century, but chemists in the eighteenth century were much concerned with understanding and improving agriculture. Earlier, in the seventeenth century, Johann Rudolph Glauber had discovered that spreading saltpeter in the soil gave a boost to plant growth. In the eighteenth century, chemists continued to investigate how and under which conditions plants thrived. Much of this work, such as Stephen Hales' and Jean Senebier's studies on plant respiration and water absorption, had theoretical value but supplied little immediate help for agricultural productivity. Most changes in farming practice came in the form of field organization, work practices, breeding, and crop cultivation, but fertilizers were also a point of interest. Some, such as manures, had long been used, but experimentation by farmers led to new ones. For example, marl, a lime-rich soil, came into more extensive use in the eighteenth century. When it was understood that lime was the most active agent in marl that promoted growth, lime burnt in kilns replaced marl soil as a fertilizer after 1760 (Clow and Clow 1952: 456–92).

More systematic agricultural chemistry emerged after 1760. Chemists such as Henry Home in Scotland and Johan Gottschalk Wallerius in Sweden began to publish experimental work. However, the variability of soil conditions, climate, and crops meant that little of this work had the scope needed to yield results of practical consequence for agriculture. At the end of the century, the emphasis in research moved from crop rotations to fertilizers, both natural and artificial. Home had already noted the fertilizing value of what is now called potassium sulfate, and Archibald Cochrane in the 1790s identified that alkaline phosphates were responsible for the fertilizing value of bone meal. Artificial fertilizers would be of great importance in the nineteenth century (Fussell 1969; Krohn 1983; Jones 2016).

In France, chemists and agronomists were interested in better understanding and improving food from 1760 onwards. Food scarcity could lead to unrest, and so the government wished to improve yields. Bread in particular was essential for keeping people from starvation. The nutritional qualities of bread were analyzed by Jacopo Bartolomeo Beccari in 1742 in Bologna. He showed that wheat had two nutritive components, one starchy and the other glutinous. Chemists debated the nutritive value of gluten and starch, and whether wheat bread was the best staple food. Other staples from outside Europe, such as potatoes, were introduced and become more important in diets, but their relative merits were unclear. Advocates of new foods, such as the pharmacist Antoine-Augustin Parmentier, promoted the wider acceptance of potatoes based on chemical and medical arguments, rather than treating bread as a simple food. He argued that starch was chemically identical in different foods such as wheat and potatoes, and that it was more important than gluten in nutrition. Although he had little success in displacing wheat bread from the French diet, the arguments that he and others made about situating the alimentary value in chemical terms was lasting. Chemical and medical arguments would thus alter the understanding of what was essential in nutrition (Spary 2014).

OTHER CHEMICAL INDUSTRIES

Alkalis

One of the most heavily used family of chemical substances of the eighteenth century was alkalis. The term includes a range of chemicals that are now recognized as alkaline metal salts and oxides. The primary alkalis were potash or the more refined pearl ash (potassium carbonate) and soda (sodium carbonate). These substances were indispensable in a wide range of uses, including the production of alum, glass, soap, and saltpeter for gunpowder, and in washing and dyeing of textiles. Potash was made by first burning wood, peat, or other plant materials in pits. The resulting ash was boiled with water in iron pots to concentrate the potassium carbonate solution, then crystallized and separated, at which point it acquired the name potash. The potash, at about 40 percent potassium carbonate, could be further refined in a reverberatory furnace into pearl ash at 80 percent purity. Forested countries, such as Sweden and Russia, were important exporters of potash. Soda, by contrast, was produced mostly in Spain by burning barilla plants or seaweed and then deriving the product by a similar process as for potash. Mineral deposits of impure soda, called natron, were also mined in Egypt. Barilla was exported throughout Europe, but other sources of soda existed, such as kelp harvested from seashores in Ireland and later in Scotland, or the saltwort plant in southern France. These sources, especially kelp, became significant after 1750, although exact production figures are not available (Clow and Clow 1952: 65–75; Smith 1979: 192–4).

Given the importance of soda for a wide variety of industries, the artificial synthesis of soda was recognized as a potentially transformative process, especially in countries that had to rely on imports. Henri-Louis Duhamel du Monceau synthesized soda in his laboratory in 1737, but his reactions were not practicable on a large scale and it did not stimulate any commercial activity. Forty years later, during the American War of Independence, the decline in imports into France of the potash needed for making gunpowder prompted the Académie Royale des Sciences to offer a prize for a method to synthesize alkali in 1781. Although this produced little of immediate practical value, it prompted research into the matter. The physician Nicolas Leblanc experimented with the synthesis of soda from 1784. In 1789 he discovered the process that was to bear his name and would serve as the basis for the soda industry for much of the nineteenth century. It involved treating ordinary salt with sulfuric acid to yield sodium sulfate, which was then reacted with charcoal and limestone to give sodium carbonate, among other products. The soda could then be isolated and purified by leaching it out of the resultant mixture.

Leblanc succeeded in convincing investors to fund a plant at Saint-Denis, north of Paris. It managed to produce soda in trials by 1793, but operations were interrupted by the upheaval of revolution and war, as the government sequestered raw materials, especially sulfur for gunpowder. Attempts to revive the project came to little until after 1800. In France the first successful soda plant was established in Grenelle in 1797 (Smith 1979: 209–48; Schranz 2014: 257).

In Britain, six patents were awarded between 1779 and 1789 for methods of producing sodium sulfate from salt and then transforming it into soda. One of the first had appeared in the 1760s after John Roebuck, James Watt, and Joseph Black, later joined by James Keir, created a process to make soda. Keir then established a soda works at Tipton, later converted to soap production. Synthetic soda production, however, remained a very small affair in Britain, as also in France, until 1823, when James Muspratt built his Liverpool alkali works based on the Leblanc process. Leblanc soda became a huge business in the course of the nineteenth century.

Gunpowder

Gunpowder is a mixture of saltpeter (potassium nitrate), sulfur, and charcoal. It had been produced in various ways since the Middle Ages, and more stable and powerful forms had emerged in the Renaissance leading to its broader adoption in weaponry. In contrast to some of the other materials covered in this chapter, its process of production underwent no significant change. It was, however, very important given the constant warfare that characterized this century. Each of the three components of gunpowder came from different sources. Charcoal was made by charring wood, while sulfur was mined, Italy being a major supplier. The sulfur had to be refined by heating it in an iron pot and straining it. Saltpeter forms naturally in soil containing rotting nitrogen-rich organic materials. It could then be leached from the soil and crystallized in various ways. By the eighteenth century, an international trade in saltpeter had grown, with India being a major exporter via the Dutch and British East India Companies.

The three ingredients of gunpowder had to be further refined, carefully mixed with a little water (to prevent inadvertent explosions) in "incorporating mills" to produce pellets ("corned" powder) that were reasonably stable. It was difficult to produce good quality gunpowder given the complex production process. Governments were also very concerned about their supplies and made efforts to secure them. They began to run their own gunpowder manufactories, such as the British Ordnance Board's gunpowder mills at Waltham Abbey, purchased in the 1780s after concerns were raised about the ballistic value of purchased supplies. In France, the government established the Régie des Poudres in 1775, naming Antoine Lavoisier as one of its governors. One of the world's largest factories in the 1790s was the British military Ichapur Gunpowder works in India, employing around 2,500 people. Governments were also interested in improving the quality of their supplies. They encouraged chemists, such as Lavoisier, and military engineers, such as William Congreve, to determine how the powder's ballistic characteristics could be improved. They made gradual improvements, typified by Lavoisier's work on determining the composition of saltpeter (Mauskopf 1995; Buchanan 2006; Mauskopf 2010).

CONCLUSION

By the end of the eighteenth century, many of the trends that would give rise to the industrialized chemical industries of the nineteenth century had already begun to emerge, albeit haltingly, and in only a small number of industries. These trends include the increased capitalization of industries as larger factories replaced artisanal workshops. The largest factories of much of this period were run or sponsored by governments, including the gunpowder mills and the royal manufactories, some dating back to the seventeenth century. By 1750, however, a number of other sites were growing in scale. These included iron smelters, early Leblanc soda plants, and the St. Rollox chemical works at the very end of the century.

Systematic research in chemistry, whether in an academic or industrial context, was another significant trend. Research and experimentation was apparent in a wide range of industries, such as in the production of sulfuric acid, zinc, pottery, and food. Some of this was based in formal academic contexts. The discovery of chlorine and expanding knowledge of gases are examples. Others seem entirely internal to the industries in question: for example, the improvements in iron puddling or the introduction of new sources of soda such as kelp. The circulation of knowledge and the commonalities in interests, however, are readily apparent in many cases.

There were other trends with deeper histories that continued to affect chemical industries in the eighteenth century. Most notable among these was the growth of international trade, which circulated goods such as textiles, dyes, and pottery around the world and stimulated chemical industries, either by giving inspiration or to be consumed directly. The steady rise over the course of the century in demand for consumer goods was also a stimulus for chemical industries in Europe, as they met demand for a wider range of goods, from printed textiles to brighter lights.

CHAPTER SEVEN

Learning and Institutions: *Didactic Chemistry and Practical Instruction*

JOHN C. POWERS

INTRODUCTION

The traditional historiography of eighteenth-century chemistry has focused on the chemical revolution, specifically the claims and practices of Antoine-Laurent Lavoisier, his supporters and his opponents. Historians of chemistry have examined and debated the nature and core of the revolution: the overthrow of phlogiston theory, the discovery of oxygen and its role in combustion and the formation of acids, the importance of establishing a mass balance between reactants and products, and the new chemical nomenclature. The volume of historical contributions on these topics has been so great that several historians have referred to a "Lavoisier industry." Ironically, these same historians question whether the characterization of eighteenth-century chemistry that emerges from these studies is valid (Donovan 1988; Mauskopf 2007).

Over the past few decades, the history of eighteenth-century chemistry has undergone a revision in its scope and emphasis. Rather than seeing pre-1770s chemistry as the foreground to Lavoisier, historians such a Frederic L. Holmes examined chemists and their work on their own terms and within the context of contemporary problems and institutions (Homes 1989). In addition, rather than focusing narrowly on changes in chemical theory or a handful of new practices, new studies, in the words of John Christie and Jan Golinski, focus on "the whole range of cultural conditions governing both practical chemistry and chemical discourse," that is, "the human activities of practicing and talking about chemistry" (Christie and Golinski 1982). The resulting picture that has emerged suggests an ongoing revolution (or, perhaps, simply an accumulation of changes), which saw chemistry develop from a craft practice into an academic, philosophical art that involved a range of chemical practitioners, a plethora of new theories and techniques, and the growth and development of new institutions (Kim 2003; Klein 2015b).

At the center of these developments was the expansion and reform of pedagogical practices and institutions. Regardless of what metric one uses, the number of chemistry courses, university professorships, textbooks, or students, education in chemistry experienced precipitous growth during the eighteenth century. This growth reflected a rise in the status of chemistry, which was increasingly seen as both useful in its application to solve technical problems, and as a field that could contribute to philosophical discourse (Meinel 1983; Principe 2007).

At the beginning of the century, most chemistry teachers were apothecaries or other artisan-chemists. For those who wished to join one of the chemical trades – pharmacy, assaying, metalworking, dye and pigment making, etc. – the traditional route through apprenticeship remained intact. However, practical training was increasingly supplemented by chemistry lectures and book study. Didactic chemistry courses and textbooks were originally created to help train apprentices and medical students, but they also became instruments to augment the reputation of lecturers and promote chemistry to a wider audience. Thus, teaching chemistry and composing textbooks became a route to social advancement for chemists, and their lectures came to be seen as essential training for medical students, natural philosophers, and those participating in a wide variety of trades and commercial applications.

My approach to the pedagogy of chemistry suggests that the act of teaching a course or writing a textbook is a form of *doing* chemistry. Didactic presentations of chemistry require the lecturer to organize and systematize a diverse array of facts, ideas, and techniques, an activity which compels him to articulate the parameters and nature of chemistry as a field of practice and inquiry. Given the limited opportunities for publication in the eighteenth century, courses and textbooks became the media through which many chemists presented novel phenomena, proposed new theories, and conducted public debates with other chemists. In addition, the textbook provided a framework into which new material could be integrated into extant knowledge. So important was this feature of didactic chemistry that Lavoisier summarized his new chemistry in textbook form, as the *Traité élémentaire de chimie* (1789), and planned to pen a more comprehensive textbook, which incorporated subjects omitted from the *Traité* (Bensaude-Vincent 1990).

The following chapter examines the growth of eighteenth-century chemical pedagogy and institutions by examining the cultural circumstances and practices of chemists-educators and students. I begin by looking at the training of apothecaries through apprenticeship, its drawbacks and attempted reforms. Then I discuss the rise of university chemistry and the creation of other institutions, such as schools of pharmacy and schools of mining. Finally, I delve into didactic chemistry and its development as a means of structuring chemical knowledge. What I hope to provide is an alternative narrative to the traditional story of eighteenth-century chemistry. I aim to tell a story about the rise of chemistry through pedagogy and its institutions, which does not look backward from the perspective of the chemical revolution, but rather places actors and institutions within the developing tends of the century.

ARTISAN CHEMISTS AND PRACTICAL TRAINING

During the eighteenth century, apothecaries and other artisan-chemists controlled the practical training of chemists. Any student who wished to learn how to practice chemistry, regardless of whether they ultimately aimed to become an artisan-chemist, an informed physician, a university professor, or an educated gentleman, needed to obtain practical training in the laboratory. Essential for this goal was hands-on knowledge of how to operate chemical apparatus and instruments, maintain the furnaces, use solvents, and master some basic operations and analytic techniques. For this training, an aspiring chemist, or his family, would have to come to an arrangement with a willing apothecary or some other chemical practitioner. The traditional route for obtaining such training, especially for those intending to enter one of the chemical trades, such as pharmacy, assaying, or metal-working, was apprenticeship. In his foundational study of the eighteenth-century German chemical community, Karl Hufbauer has argued that about half of the elite German chemists, whom he defined as those who made significant contributions to chemical knowledge or education, were trained as apothecaries through apprenticeship (Hufbauer 1982: 55). This process, however, could be long and difficult. In Germany, for example, boys became apprenticed to apothecaries in their early teens and typically served for five or six years, followed by six to eight years as a journeyman. Young apprentices were charged with menial tasks such as cleaning utensils, maintaining the chemical furnaces, selling prepared medicaments in the dispensary, and drying herbs. They were not usually allowed to learn how to prepare medicines, that is, to undertake proper "chemical" work, until their fourth year of apprenticeship (Kremers and Urdang 1976: 92–3; Klein 2007: 99–100).

There were other limitations to traditional apprenticeship. The quality and kind of education that an apprentice received depended greatly on where the young man apprenticed and who the master apothecary was. Some apothecaries, such as those in more rural areas, might not have a fully equipped laboratory. Even in better-equipped establishments, the training of apothecaries was often delegated to journeymen or other workers in the shop, who assigned to the apprentices the least-desirable tasks and often took little interest in their educational progress. A widespread complaint made by apprentices was that their work was boring and repetitive. The chemical training that they received was practical: how to make medicaments and other chemical products. They were expected to learn procedures by observing more experienced workers. Once an apprentice had mastered a specific process, he would typically be called on to perform this process as often as needed (Klein 2007: 101–9). There were, of course, examples of master apothecaries who spent more time to educate apprentices beyond this basic level. In one anonymous 1787 account from Germany, the master himself taught the younger apprentices the proper terminology for pharmaceutical simples and Galenic *composita*, introduced them to affinity theory and other "chemical principles," had them read a recently published pharmacopoeia and handbook, and even required them to practice their Latin. Older apprentices were expected to study more advanced chemistry texts, including articles from Lorenz Crell's recently founded journal, *Chemische Annalen* (Klein 2007: 110–12).

As this last example suggests, over the course of the eighteenth century there were efforts to reform the training that apprentices received. These efforts were driven by the fact that in most places in Europe, the practice of apothecaries was regulated. All novice apothecaries had to take a licensing exam administered by the local medical authorities, usually the local medical college or apothecaries' corporation or guild (see Kremers and Urdang 1976: 63, 76–7, 92). In preparation for these exams, regulatory institutions and, also, the occasional apothecary master would provide additional instruction to novices, usually in the form of chemistry courses. For example, in response to a 1725 Prussian royal edict, which stated that apothecaries had to pass an exam in practical chemistry administered by the *Collegium Medico-Chirurgicum*, the *Collegium* in Berlin established regular chemistry courses taught by Caspar Neumann and Johann Heinrich Pott to review the preparation of remedies and their practical uses (Hufbauer 1982: 55–6; Klein 2007: 100–1). In France, courses for apprentice apothecaries had been mandated since the sixteenth century (Kremers and Urdang 1976: 76). However, regular instruction did not begin until the seventeenth century with courses taught by private apothecaries, such as Jean Beguin and Nicolas Lémery, and through sponsored institutions, such as the public courses offered at the Jardin des Plantes in Paris. (These didactic chemistry courses are discussed below.)

In addition, chemistry itself underwent dramatic changes in its aims and status, and this shift shaped the teaching of chemistry and pharmacy. In the eighteenth century, chemistry as practiced by elite chemists became a form of experimental philosophy, in which making new knowledge and ordering and debating facts and theoretical claims was a paramount concern. In France the Académie Royale des Sciences institutionalized this new chemistry by creating a section for chemistry in 1699, which generated patronage and social status for chemists, who were willing to investigate chemical problems and engage in theoretical debate (Holmes 1989; Principe 2007). Jonathan Simon has argued that in Paris in the 1770s apothecaries embraced theoretically informed chemistry at the expense of grubby practicality as part of their drive to professionalize and redefine themselves as "pharmacists" (2005). This shift can be seen in their chemistry courses. Whereas the courses of Lémery included a discussion of the medicinal uses of each preparation he described, Antoine Baumé, the docent of the new *Collège de Pharmacie* (founded 1777), omitted this information from his monumental textbook. Claiming that one's practice was grounded in abstract, theoretical knowledge conferred academic status (Simon 2014). As with the Parisian pharmacists, in Berlin apothecaries impressed by the courses of Neuman, Pott, and then Sigismund Marggraf began to see the advantages of acquiring broader chemical knowledge: more efficient processes, purer and (therefore) better drugs, and best of all, access to the social circles of higher learning and culture (Hufbauer 1982: 55–7).

A chemical novice enduring the traditional apprenticeship was often limited in his access to this new chemistry. If an apprentice desired to expand his knowledge by delving into chemical philosophy or concepts – to try to understand *why* substances behaved the way they did – he often had to seek access to books or other sources outside of the apothecary's shop. For example, during his apprenticeship (1748–54) under the Dresden apothecary C.F. Sartorius, Johann Christian Wiegleb desired to know the "reason and conception" behind his work, so he read the textbooks he found in the small library at his master's shop. Only later did he realize that the texts, all published during the seventeenth century, were very out of date, and so he had to seek out contemporary sources (Klein 2007: 101–2). In university towns with medical schools, it was not uncommon for apothecaries to attend university lectures on chemistry or other medical subjects. In Edinburgh, where both surgeons and apothecaries were member of the Incorporation of Surgeons, apprentices made up the majority of students in university chemistry courses during the first three decades of the century. In many instances, apprentices demanded or expected to be allowed to attend university courses, believing not only that this improved their education, but made their training and profession more "genteel" (Rosner 1991: chap. 5). Even as medical students became the majority in Edinburgh chemistry courses, William Cullen and Joseph Black still attracted "industrious apprentices" throughout their tenure (Anderson 2010: 86–7).

By the closing decades of the eighteenth century, many apothecary-chemists, dissatisfied with the traditional system of apprenticeship, sought to reform the training of apothecaries to include both theoretical knowledge and a wider range of practical knowledge. As Ursula Klein has pointed out, even though the books in Sartorius' library were out of date by the 1750s, the fact that he had a library at all speaks to the rising status of apothecaries in German society. The most successful were prominent merchants and purveyors of culture and learning, and these elite apothecary-chemists participated in the development and dissemination of chemical knowledge (Klein 2007). Several of these men founded institutions aimed at reforming the training of apothecaries. In 1779 Wiegleb opened a chemical boarding school in Langensalza, as did Johann Trommsdorff in Erfurt (1795) and Sigismund Hermbstädt in Berlin (1789). All of these schools advertised the teaching of "chemistry in its entirety," which for Hermbstädt's school included physics, mineralogy, pharmacy, assaying, and metallurgy (Hufbauer 1982: 190–1, 210–12, 218–19; Klein 2007: 110–13).

The elite apothecary-chemists identified by Hufbauer received their training from a variety of sources. Although the German chemical boarding schools were innovative for their time, their impact, when one considers the entirety of the eighteenth century, was limited. In most cases, ambitious chemical novices sought out books and courses to take, but they also traveled and worked in a variety of laboratories, learning what they could from a variety of mentors. In Germany, it was not uncommon for apothecary-chemists to work in five or more establishments as journeymen before settling in one place. Before Martin Klaproth established his own shop in Berlin, and eventually became a chemical lecturer, renowned researcher, and first full professor of chemistry at the University of Berlin (1810), he worked as a journeyman at the *Ratsapotheke* in Quedlinburg and in Danzig, at the *Hofapotheke* in Hanover, and in two private establishments in Berlin (Dann 1958; Klein 2007: 125–6).

Chemists who had family resources or patronage could expand their travel beyond apothecary shops and gain knowledge and connections from physicians, philosophers, and other chemical artisans. French apothecary-chemist Etienne-François Geoffroy, the creator of the first affinity table, initially labored in his father's apothecary shop in Paris, before traveling to Montpellier to complete his training in pharmacy. He then proceeded to take a grand tour, traveling along the French seacoasts, then to Holland, England, and Italy, meeting the savants and "illustrious men" of those countries. His experience and connections made on this tour recommended him as an ideal choice for election to the Académie Royale des Sciences in 1699 (Joly 2014: 118–19). In another example, Caspar Neumann, after serving an apprenticeship in his hometown of Züllichau and working in Berlin as a journeyman, became a journeyman for the Royal Travelling Apothecary to King Frederick I of Prussia. He negotiated this post into a source of royal patronage, persuading the court to fund his travel to the

Harz Mountains and other mining regions to learn metallurgy and assaying; to tour glassworks, foundries, and botanical gardens; and ultimately to travel abroad to Holland. After the death of King Frederick in 1713, Neumann ran the London chemical laboratory of the Dutch surgeon Abraham Cyprianus, until he was persuaded by Georg Ernst Stahl to re-enter Prussian royal service in 1716. The court funded further travel for him to Paris and London until 1719, when he finally settled in Berlin as a royal apothecary (Hufbauer 1982: 173–4; Klein 2007: 116–17).

THE RISE OF UNIVERSITY CHEMISTS

Over the course of the eighteenth century chemistry gained a foothold in European universities. At the beginning of the century there were only a handful of university professorships of chemistry, and many of these were not regular or sustained positions. Chemistry was considered to be an extracurricular subject, which most university faculty saw as having only a minor role in medical training. This situation changed during the course of the century as the status and perceived value of chemistry increased, and university faculty and administrators sought to integrate it into the curriculum. There were only six salaried professorships of chemistry in German medical faculties in 1720; by 1780 there were twenty-eight (Hufbauer 1982: 34). This increase reflects not only the increased standing of chemistry within medicine, but also the value placed on chemistry and chemical analysis for other fields, such as natural history, mineralogy, and practical trades, such as metallurgy, pottery, and glass-making (Golinski 1992; Eddy 2007; Eddy 2008).

The incorporation of chemistry into the university curriculum over the span of the eighteenth century was haphazard and depended upon local circumstances. For much of this period, chemistry occupied a marginal position in most university curricula. In the minds of most medical faculty professors, chemistry as pharmacy or *materia medica* was practical knowledge which fell under the purview of the apothecary, a tradesman. At best, chemistry was an auxiliary or "handmaiden" to medicine and was not required for a medical degree (Debus 2001). Whereas the university physician claimed his status through his mastery of theoretical knowledge and book-learning, chemistry was dirty and merely practical. In his inaugural oration as professor of chemistry at the University of Leiden in 1731, Hieronymus David Gaubius acknowledged this perception by imploring his students to leave their writing desks and come sweat with him in his laboratory, where they could blacken their hands with smoke, soot, and cinders (Meinel 1988: 89). Despite Gaubius' urgings, relatively few university students worked in their professor's laboratory. More typically, professors or would-be professors of chemistry gave didactic chemistry courses, in which students heard lectures and observed processes being demonstrated.

As Christoph Meinel has pointed out, for many university professors teaching chemistry courses was a stepping stone to a more prestigious medical chair or as means to gain an appointment or patronage, usually as a physician, outside of the university (1988: 93–4). In almost all cases professors of chemistry who were also physicians had other duties and interests apart from chemistry, such as teaching core medical courses or botany, attending to patients, and performing administrative duties.

In university towns with medical faculties, apothecaries and other artisan-chemists were often the first to offer didactic courses and other training to university students. Many of these apothecary-chemists became the first professors of chemistry at their institution. In effect, for those outside of the university, giving chemistry lectures became a route for advancement into academe. For example, in 1683 the Veronese apothecary Giovanni Francesco Vigani began giving private chemistry courses in Cambridge, and he had already published a short textbook, *Medulla chemiae* (1682). His popularity among his students enticed the University Senate in 1703 to grant him the title of "Professor of Chemistry," albeit without a salary or specific duties. Nevertheless, Vigani's success at attracting students and others to his courses solidified his professorship as a stable position, which the university felt obliged to fill upon his death in 1713 (Guerrini 1994: 186–8; Schaffer and Stewart 2005). Similarly, Johann Conrad Barchusen, who was a practicing apothecary, convinced the University of Utrecht in 1703 to make him professor of chemistry after he had given private lectures for nine years and published textbooks in both pharmacy and chemistry (Hannaway 1975). Even in Leiden, the first two professors of chemistry, Caerl de Maets and Jacob le Mort, were trained as apothecaries and had to prove their competence and ability to attract students before the university curators granted them salaried positions. Only after the appointment of Herman Boerhaave in 1718 did university-trained physician-chemists become the norm for Leiden's professors of chemistry (Van Spronsel 1975; Powers 2012).

University students who wished to learn the practical skills needed to work in a laboratory typically sought training from practitioners outside of the university, such as local apothecaries or in the private laboratories of other chemists. The primary tool of chemical instruction at the university was the didactic course, which professors and administrators deemed sufficient for most university students, who only needed to know about chemistry to inform their medical practice or commercial projects. As physicians or educated gentlemen, they were most likely never going to practice chemistry. However, for the smaller group of students who wanted to acquire practical skills, hands-on training was required. Occasionally, an enthusiastic medical student who cultivated a relationship with a professor practiced in chemistry received manual instruction and the opportunity to work in the university's or the professor's private laboratory. In Glasgow, for example, William Cullen employed his student Joseph Black as a laboratory assistant in 1748, and there are other instances in which medical students experimented with their professors, usually within the context of composing a chemistry-related medical thesis (Anderson 1978: 10, 19, 27). More commonly, students who could pay or trade on other resources made arrangements with local apothecaries for private lessons or for the opportunity to work in their laboratories. While he was a student at the University of Leiden, Boerhaave received practical training in chemistry by working from 1691 to 1693 in the laboratory of a local apothecary, David Stam (Lindeboom 1973; Powers 2012: 57–9). In other centers of medical study, such as Paris and Edinburgh, there were always apothecaries or other chemical practitioners who were eager make similar arrangements with medical students for such training (Anderson 2010; Perkins 2010: 41–3). Some students could also trade on their labor, skills, or status to gain access to practical training. In 1779, Henri-Albert Gosse negotiated to spend six months in the Parisian laboratory of apothecary Antoine Arnaud Quinquet, in return for his labor in assisting with pharmaceutical preparations. He then negotiated a further six months acting as an assistant in Antoine-François Fourcroy's private laboratory (Perkins 2010: 42).

Despite the ambiguous connotations associated with chemistry within the university, these institutions helped to increase the status of chemistry over the course of the century. Some university medical faculties, such as Jena and Leiden, recognized the importance of chemistry for medicine and established professorships in the seventeenth century. By the turn of the eighteenth century, the medical faculty at Leiden, for example, had a well-established tradition of education and experimentation in physiology and anatomy, which included the chemical analysis of bodily fluids and tissues as well as medicaments (Ragland 2008; Powers 2012; Ragland 2017). To support the demand for chemistry, the curators of Leiden established a chemical laboratory and professorship in chemistry in 1669 (Van Spronsen 1975; Powers 2012: 47–51). The medical faculty at the University of Jena founded a professorship in 1639 first held by Werner Rolfinck, who integrated chemistry into the medical curriculum and included chemical topics in the faculty's academic disputation exercises (Partington 1961: 312–14). Graduates of Jena and Leiden then brought this chemically inclusive medicine with them as they became medical practitioners and faculty at other institutions. Jena students Georg Ernst Stahl and Friedrich Hoffmann established chemistry courses at the University of Halle, as did Johannes Bohn at the University of Leipzig (Partington 1961: 653–86, 690–700; Hufbauer 1982: 165, 167–9). The students of Herman Boerhaave in Leiden similarly sought to establish chairs of chemistry at other institutions. Gerard van Swieten, for example, established a chair at Vienna in 1749, and as an official of the Hapsburg Court, he attempted to found chairs at other institutions throughout Hapsburg territories, but with limited success (Hufbauer 1982: 266; Meinel 1988: 95–6). In the 1720s Leiden students more successfully transplanted the Leiden model to Edinburgh. Each of four founders of the medical faculty, including Andrew Plummer, who primarily taught the chemistry courses, had studied with Boerhaave (Anderson 2006; Powers 2015).

A second and in some areas greater factor for the growth of chemistry lectureships and professorships was the increasing importance for chemistry within practical trades and other commercial applications. In Germany, this made chemistry a valuable body of knowledge for state administrators, tradesman, and other gentlemen who wished to promote modernization projects (Meinel 1988: 97–8; Klein 2012c). These men attended university chemistry courses as well as those at the various technical schools, such as the Mining Academy at Schemnitz. In Sweden, mining officials, engineers, and chemists attended chemistry courses at universities, such as those offered from 1750 by Johan Gottschalk Wallerius at Uppsala, and similar courses begun at Lund (1758) and Åbo (1761). Notably, these new professorships in chemistry were located in the respective university's philosophy faculty rather than the medical faculty, indicating the emergence of chemistry from strictly medical applications in European university curricula (Meinel 1988: 98; Fors 2015: 93–6). Even at the University of Edinburgh, where chemistry continued to be taught in the medical faculty throughout the century, a significant portion of the students attending the chemistry courses of Cullen (between 1755 and 1766) and Black (between 1766 and 1797) never studied medicine, but rather their interests were in trade, manufacture, or natural philosophy (Golinski 1992; Anderson 2006). Others studied chemistry at Edinburgh within the context of natural history, notably geology, in which field the chemical analysis of minerals was becoming a standard practice (Eddy 2007).

Taking a closer look at the educational paths of chemists shows many similarities among those who became, using Hufbauer's term, elite chemists. Regardless of whether the chemist was initially trained as an apothecary or was a university medical student, all received some form of practical training, combined with chemistry lectures and book learning. This pattern also applies to chemists who were neither physicians nor apothecaries nor mining chemists. Antoine-Laurent Lavoisier, the Parisian lawyer turned chemist who was the central figure in the chemical revolution of the 1770s through 1790s, was introduced to chemistry in 1762–1763 through the boisterous lectures of Guillaume-François Rouelle at the Jardin du Roi. Henry Guerlac suggested that Lavoisier may have also worked in Rouelle's apothecary shop. He gained additional practical experience though his work with Jean-Etienne Guettard, a mineralogist, who took the young Lavoisier on geological surveys, during which he would perform chemical analysis on minerals (Guerlac 1956).

DIDACTIC CHEMISTRY: AUDIENCES AND INSTITUTIONS

By the eighteenth century the most common form of chemical education was the didactic lecture. Apothecaries and other artisan-chemists could augment their incomes by offering formal chemistry courses and, in some cases, penning textbooks or handbooks in chemistry, many of which sold quite well. These courses and textbooks became an important part of chemical education for artisan-chemists, physicians, and members of the general public who had an interest in chemical matters. Originally conceived to help apothecaries pass their licensing exams, for some skilled lecturers presenting courses and writing textbooks became a major source of income and a route to help them gain access to academic and higher social circles.

The first chemistry courses were designed to help novice apothecaries learn the basic chemical terms, concepts, and techniques so that they could pass their licensing exams. The Parisian apothecary Jean Beguin offered his chemistry course to apprentices around the turn of the seventeenth century for this purpose. To supplement his course, in 1610 he published a short textbook, *Tyrocinium chymium*, in which he cribbed heavily from what may be considered the first textbook of chemistry, Andreas Libavius' *Alchemia* (1597) (Clericuzio 2006: 240–3). Nevertheless, Beguin's book and course spawned a succession of didactic chemistry courses and textbooks. The appeal of Beguin's course was both the presentation of chemical knowledge in an organized fashion and the fact that he included descriptions of controversial, Paracelsian, or "chemical" remedies in his text. These two things promoted Beguin's book and the method of organizing and teaching chemical knowledge that it contained. Pirated editions of the *Tyrocinium* appeared soon after he published the first edition, and subsequent chemist-publishers changed and added material to Beguin's framework to suit their own interests. For example, a 1625 edition published in Marburg was already three times the size of Beguin's original text (Patterson 1937). Over the course of the seventeenth century, more apothecaries taught courses out of their shops, but patrons of the new learning also created sponsored lectures. In 1648 the French physician Guy de la Brosse founded the best-known sponsored lectures at the Jardin du Roi in Paris. These lectures were given by a paid *intendant*, usually an apothecary, for the purpose of instructing apothecaries and physicians in chemical remedies, and continued through the eighteenth century (Contant 1952; Howard 1981; Clericuzio 2006: 343–5).

By the early eighteenth century the audience for chemistry courses had grown far beyond apothecaries and apprentices, as chemistry had become a topic of interest for the cultural elite, and in some places chemistry courses were a form of entertainment. For example, by the beginning of the eighteenth century, Lémery was the most successful chemistry lecturer in Paris (Bougard

1999; Clericuzio 2006: 347–50). His courses regularly attracted large audiences composed of artisans, physicians, philosophers, and interested gentlemen and ladies "swept away by fashion," as Bernard Fontenelle, secretary of the Académie Royale des Sciences, described them (Fontenelle 1719). Reflecting the initial aims of the didactic tradition, Lémery's courses focused on practicality and utility: defining chemical terms and concepts and demonstrating chemical recipes, which always included a discussion of the recipe's uses. He would, however, tailor his courses to fit the interests of his diverse audience, often including some demonstrations simply for their novelty. In his popular courses, for example, he demonstrated the action of burning lenses and the ability of phosphorus to generate "artificial light," operations of no practical value at the time, but included in order to elicit a sense of wonder (Lémery 1683). Lémery published a textbook, *Cours de chymie*, which he expanded and amended over fifteen editions from 1675 through 1715. The tenth edition of the text (1713) ran to over 900 pages (Lémery 1713). Despite its length and technical nature, Fontenelle commented that the book "sold like a work of romance or satire" (Fontenelle 1719).

Lémery's success demonstrated that didactic chemistry was well established and commanded a large and diverse audience. This audience and popularity continued to grow during the course of the eighteenth century. John Perkins has estimated based on extant records that in 1786 just over 3,000 individuals attended various chemistry courses in Paris, a threefold increase from 1761, and probably a fivefold increase from Lémery's day (Perkins 2010). The audience for these courses remained diverse as well. Rouelle held the title of "professor and demonstrator" of chemistry at the Jardin du Roi, where he presented public courses from 1743 until 1768. His courses attracted enormous crowds, which often overflowed the 600-seat auditorium at the Jardin. His audience, like Lémery's, was comprised of a diverse swath of Parisian society, including members of the nobility, governmental administrators, engineers and army officers, medical and pharmacy students, physicians who came to Paris to continue their education, and students at the Ecole des Mines and Ecole des Poudres. Notable students of Rouelle were the chemists Pierre-Joseph Macquer, Gabriel-François Venel, and Antoine-Laurent Lavoisier, and the *philosophes* Denis Diderot and Anne Robert Jacques Turgot (Bensaude-Vincent and Lehman 2007: 83–5; Roberts 2008; Perkins 2010: 37).

Perkins has established a useful taxonomy of eighteenth-century chemistry courses, which speaks to the diversity of available courses and potential students. Institutional courses were those which were sponsored by an institution, where students of the institution or on occasion the general public may attend without paying lecture fees. The public courses at the Jardin du Roi would fall into this category, as would university courses which were open to all matriculated students. *Private* courses were those which an instructor offered to anyone who was willing to pay a lecture fee. Most courses offered by apothecaries in their shops fall into this category. These courses would often be advertised with posters, in medical or pharmacy journals, and newspapers, depending on the intended audience (Bensaude-Vincent and Lehman 2007: 84). Individual courses were not open to the public, but rather were negotiated between an instructor and an individual or small group, and may have included more hands-on instruction that would not be possible in a larger course (Perkins 2010: 28). Chemistry instructors often offered several kinds of course during their careers, and often at the same time. For example, from 1718 to 1728 Boerhaave as professor of chemistry at the University of Leiden fulfilled his obligation to teach one institutional course that was open to all matriculated students by lecturing on the "chemical instruments." Simultaneously, he offered a private, term-length review of chemical theory and practice, arguably the more useful course for most students and for which he collected lecture fees (Powers 2012).

Most lectures early in the century were still held in the dispensaries of apothecaries, but the institutions which hosted lectures increased over the century. As discussed earlier, universities began creating chairs of chemistry, and other institutions that had a stake in training or regulating medical or artisanal practice began to offer didactic courses as well. By the 1780s in addition to private courses and the public course at the Jardin du Roi, a Parisian student could take regular chemistry courses at the Collège de Pharmacie, the Collège de France, the Faculté de Médecine (after 1785), the Collège de Chirurgie, or the Royal Mint (Perkins 2010: 35–6). The popularity of didactic chemistry also extended to the French provinces, where institutions such as the Faculty of Medicine in Nancy and the Société d'Etude des Sciences et des Arts in Metz (founded 1757) regularly offered public courses (Perkins 2003; Perkins 2004). Similarly, the lectures of Pott and Neuman at the Collegium Medico-Chirurgicum in Berlin and at the new chemical boarding schools followed this pattern (Klein 2007: 100–1). Newly established mining academies in Germany often combined didactic courses with practical instruction and fieldwork. The Academy at Schemnitz in the Habsburg Empire (now Banská Štiavnica, Slovakia) was established in 1763 and became the main educational institution for those who wished to become mining officials in Habsburg territories. Anton von Ruprecht became Professor of the Chemistry of Mining in 1779, and taught chemistry and mineralogy through didactic lectures and practical instruction in workshops (Konecný 2012: 336–40). Similarly, at the Swedish Bureau of Mines, Georg Brant trained mining chemists at the Bureau's *laboratorium chemicum*. While much of this training was practical, lecture notes from didactic courses taught by Brandt survive. After the University of Uppsala appointed its first professor of chemistry, Johan Gottschalk Wallerius, most Swedish students who were to become mining officials attended chemistry courses at the university and then completed their practical training at the Bureau (Fors 2015: 93–6).

While the number of institutions that offered instruction in chemistry proliferated in the eighteenth century, the number of lecturers without an institutional affiliation or connection to an apothecary's business also increased. With the growing popularity of chemistry, one could make a modest living as a chemical lecturer and publisher, but as with lecturers on other subjects, the primary goal in doing this was to secure a permanent position or aristocratic patronage. In England, the most successful of these lecturers was Peter Shaw, who made his living by editing, translating, and publishing the work of other chemists and philosophers, most famously the chemistry textbooks of Boerhaave and Stahl and an edition of Francis Bacon's *Works*. He also presented his own chemistry lectures in London (1731–1732) and then in the spa town of Scarborough (1732). He did this, he claimed, to promote the utility of chemistry and to improve medicine and the trades, although as Golinski pointed out each of his publications and courses was aimed at garnering a specific kind of patronage. His strategy succeeded; he was appointed physician-in-ordinary to King George II and subsequently to King George III (Gibbs 1951; Golinski 1983). One did not necessarily need any certification or degree in chemistry to offer courses. Robert Anderson has pointed out that in Edinburgh some lecturers had no credentials (Anderson 2010). In Paris, Balthazar-Georges Sage taught chemistry at his family's apothecary shop in the 1760s, and then again at the Mint (1778) and the Ecole des Mines (1783), even though he neither qualified as an apothecary nor earned a university degree (Todericu 1984).

EVOLVING PEDAGOGICAL METHODS AND MEANINGS

Chemistry courses and textbooks in the eighteenth century did more than just communicate or disseminate established knowledge. Pedagogy has its own history, which fundamentally shapes the particular subject being taught. The primary function of a chemistry course or textbook is to categorize and organize the diverse morass of substances, instruments, phenomena, and operations associated with chemical work into a form which can be easily communicated to novices. How a particular chemist organizes all of this material, and chooses what to include or exclude, indicates much about their understanding of their field. Thus, chemistry courses and textbooks define the parameters and content of chemistry, relate how one pursues the practice of chemistry, and suggest where chemistry fits in relation to other scientific fields. Additionally, in the eighteenth century, when opportunities to publish one's work or opinions were few save for well-placed chemists, courses and textbooks also became one of the vehicles through which chemists elaborated novel discoveries, situated their work in relation to other chemists, and conducted professional debates (Meinel 1988).

In effect, courses and textbooks performed important communal functions for chemists through which they presented their views to other chemists and to the general public.

Traditional didactic presentations of chemistry almost always began with a theoretical overview of the field, followed by an ordered presentation of chemical operations and products. In the 10th edition of Nicolas Lémery's *Cours de chymie* (1713), the theoretical discussion, in which he presented the definition of chemistry, the system of chemical principles he employed (salt, sulfur, mercury, phlegm, and earth), laboratory instruments, and a glossary of terms, was relatively short (68 pages) compared to the part of the text devoted to operations (880 pages). This division of space reflects that fact that Lémery's chemistry was overwhelmingly focused on practical application: how to conduct operations to make chemical products, knowledge that a practicing apothecary or physician would need to know. However, because his Parisian audience was diverse, Lémery also included, in his own words, philosophical "reasonings" for his operations; that is, he pointed out interesting phenomena and presented theoretical explanations that were not directly relevant to the execution of the recipe itself. For example, in his discussion of the "calcination of tin," he included remarks on the fact that the weight of the tin calx was more than the original tin, a phenomenon which he attributed to the presence of "fire particles" or the broken points of acid particles in the calx (Lémery 1713). Lémery's didactic presentation of chemistry preserved the traditional role of the didactic course – the training of apothecaries and other interested parties in practical chemistry – while moving his course in a direction to make it "philosophical" and engage in dialogue with the physicians and natural philosophers of Paris.

Not all chemists agreed with Lémery's approach. The London apothecary George Wilson modeled his course on Lémery's, but he also used his course to criticize Lémery for too much philosophical speculation. The order in which he organized and presented his operations, beginning with metals, then minerals, salts, plant substances ("vegetables"), and animal substances, mirrored that of the French chemist. However, in the preface to his *Compleat Course of Chymistry* (1709), Wilson asserted that chemistry should focus on the preparation of medicines and should not concern itself with wonders or theoretical "reasonings." Instead, he rejected the "frivolous and useless trifles of other Chymical Authors," and specifically "Lémery's pompous way of Philosophyzing upon the Processes" (Wilson 1709: preface). By way of comparison, Wilson's description of the "calcination of tin" comprised one short paragraph that described how to perform the operation with no mention of the weight gain or other "philosophical" aspects (Wilson 1709: 27). Throughout the eighteenth century there were practically minded chemists who, like Wilson, rejected the incorporation of "philosophy" into chemistry, preferring to stick to recipes and empirical methods.

However, over the course of the eighteenth century chemical lecturers and textbook authors increasingly embraced theoretical or philosophical approaches to chemistry. As indicated earlier, this was partly due to the opportunities for social and professional advancement that embracing a philosophical approach generated. The success of the experimental work of Robert Boyle, Georg Ernst Stahl, and the chemists at the Académie Royale des Sciences provided models for pedagogical innovation. At the University of Leiden, Herman Boerhaave's lectures (1718–1728) on what he called the "chemical instruments" (fire, air, water, earth, and chemical menstrua) presented theoretical knowledge about chemical substances and processes through demonstration-experiments that were designed to reveal latent chemical properties as natural principles. Thus, he presented his theory of chemistry as an ordered series of experiments, for example, on the action of heat (or fire), on the generation of "airs," and on solubility of various salts, which he performed and interpreted for his students (Love 1974; Powers 2012). These lectures became the core of the theory section of his extremely popular textbook, *Elementa chemiae* (1732), and became a model for other chemical lecturers.

By mid-century, many lecturers and textbook authors integrated discussion of chemical theories and presentation of demonstration-experiments (or descriptions of them) in their courses. Next to Boerhaave, Stahl was the most influential chemist within this new philosophical chemistry. Stahl's textbook, *Fundamenta chemiae* (1723), which Shaw translated as *Philosophical Principles of Universal Chemistry* (1730), was modestly successful, but his experimental work on fermentation, from which he developed the concept of phlogiston, was extremely influential (Oldroyd 1973; Chang 2002). Rouelle in Paris and Venel in Montpellier both performed demonstration-experiments as well as traditional recipes in their courses, and they taught a theory of chemical elements (fire, air, water, earth) that also functioned as instruments, derived from Boerhaave's work, but which also incorporated important aspects of Stahl's theory of matter and phlogiston (Lehman 2009b; Lehman 2010). In the 1770s, apothecaries like Antoine Baumé presented chemistry in his course as a synthesis of Boerhaave and Stahl, embracing the phlogiston theory and deploying demonstration-experiments to convey theoretical principles. As indicated earlier, this incorporation of theory lent credibility to the professionalization strategy of French apothecaries (Simon 2014).

Another significant incorporation of theory into chemistry courses was the introduction of affinity tables and diagrams. Geoffroy devised the first "table des rapports" in 1718 based on his own observations regarding the displacement order of acids, alkalis, and metals in salts. In his *memoir*, he stated that the table was meant to be a heuristic device to help students learn the order of displacement for common chemical substances (Duncan 1996). By the middle of the century, both Cullen in Edinburgh and Rouelle and Macquer in Paris regularly used these tables in their courses as pedagogical tools. Chemical lecturers, however, developed these affinity tables, and the concept of elective affinity which was their foundation, into a core theoretical construct, which they used to organize chemical facts and demarcate chemistry from other experimental practices. Cullen, for example, asserted that true chemical changes were those that were accomplished only by fire or forces of affinity (Taylor 2008). Similarly, Macquer made affinity a central concept in his course, in effect structuring his course around the substances found in his affinity table (Anderson 1984; Lehman 2010). Joseph Black taught his students the complex affinity relations in double displacement reactions by devising a system of circlet diagrams, representing the components of the salts involved. In addition, he utilized chiasmic (i.e. X-shaped) diagrams to represent the forces of affinity in these reactions (Eddy 2014).

Some chemists also utilized their courses to present new theories and the results of their research. The best example of this is Joseph Black and his discovery of latent heat. While the professor of chemistry at Glasgow, Black conducted a series of experiments in which he observed how water, which was made to boil by lowering the air pressure of the vessel in which it was contained, absorbed a great deal of heat. He presented his theory from this observation, that the water absorbs or releases heat (or fire) to change state, in his chemistry lectures of 1757–1758. As he worked out the details of his theory through experimentation, he updated his lectures, such that his notes remained the only complete account of his work until Robison's publication of Black's notes as a textbook (Black 1803; Guerlac 1982).

When Lavoisier and his collaborators began their work in the 1770s and developed their antiphlogistic chemistry by 1785, the tradition of defining and integrating new discoveries into chemistry through a didactic presentation was long established. In his *Traité élémentaire de chimie* (1789) Lavoisier elaborated his new discoveries in didactic form. In collaboration with Jean Baptiste Michel Bucquet, he planned to compose a larger textbook, which would include all of the material from the *Traité*, plus the pair planned to perform all of the foundational experiments in chemistry, so that the new chemistry would be completely grounded in experimental facts (Bensaude-Vincent 1990). Bucquet's untimely death in 1780 stalled the project, and Lavoisier never got back to it before his own death. Thus, the integration of the new chemistry into the extant body of chemical knowledge fell to Lavoisier's other collaborators, notably Antoine Fourcroy. Despite his adherence to the new chemistry, his *Système des connaissances chimiques* (1801–1802) resorted to the traditional textbook structure, not the one Lavoisier presented in the *Traité* (Bensaude-Vincent 1990).

CONCLUSION

Throughout the eighteenth century, all elite chemists, whether they trained to become apothecaries, physicians, or for some other profession, learned chemistry through a mix of didactic courses, book study, and practical experience in a laboratory or laboratory-like setting. Over the course of the century, institutions for chemical instruction multiplied as the popularity of public lectures grew, university professorships increased, and new institutions, such as chemical boarding schools and schools of mining and pharmacy, were founded. This growth reflects the increasing status of chemistry as both a practical discipline, useful for medicine and industry, and as a form of experimental philosophy that could provide theoretical explanations for material changes. Although chemical education always focused on recipes and practical operations, increasingly over the century the theory of chemistry (i.e. philosophical chemistry) and the methods of making, evaluating, and organizing new knowledge became an important part of chemical training as well. By the end of the century, the European chemical community had an educational infrastructure that was well developed and a solid pedagogical tradition that was geared toward both practical application and experimental development.

CHAPTER EIGHT

Art and Representation: *Cultural Modalities of Chemistry in the Eighteenth Century*

JOHN R.R. CHRISTIE

INTRODUCTION

Representations, of chemistry or anything else, are of many kinds: artistic, visual, verbal, public, private. During the eighteenth century chemists both produced and and found themselves subject to a diverse set of representations. As we will shortly see, just one specific, annually delivered lecture had the capacity to provoke a range of contrasting representations, which in turn serve to provide a sense of the varying territories on which representational history may find itself: vindicatory polemic, historical memory, chemical self-representation, satirical caricature, or harsh private judgement in contradiction to public respect and philosophical dignity. In terms of visual and verbal cultures, there was also the conventional association of the chemist with a characteristic workplace and its associated experimental instrumentation, be it lecture theater or laboratory. Taking up the theme of self-representation, the following chapter begins with a sketch of the kinds of representations of their science produced by chemists themselves. It proceeds with analysis of examples of representations of chemical laboratories, then turns to further material from the visual cultures of the eighteenth century.

The persistence of representations of alchemy is noted, and possible reasons for this persistence suggested, before moving to more extended consideration of the greatest painting of any chemist, Jacques-Louis David's portrait of Antoine-Laurent Lavoisier and Marie Lavoisier. The examination of visual representations concludes with a survey of politicized representations of the 1790s, when the coincidence of political revolution in France with chemical revolution in science produced representations of chemists and their science within the popular and accessible cultural politics of satirical caricature. Other sorts of representations, of Joseph Priestley by the poet Anna Letitia Barbauld, and the deployment of central chemical concepts, by Goethe's novel *Elective Affinities*, will further demonstrate the ways in which the cultural figuration of chemists and chemistry was by no means confined to visual culture, but also had notable presence in literary representations.

CHEMISTRY AND REPRESENTATION

What follows are three representations from a series of annual chemical events from late in the eighteenth century, lectures given each year as part of the chemistry lecture course by Joseph Black, professor of chemistry at the University of Edinburgh. The particular lecture portrayed was about "fixed air," its properties and its role with respect to discriminating the different properties of mild and caustic alkalis. Fixed air is a gas (now known as carbon dioxide), discovered and chemically identified by Black in the 1750s, a discovery which made his reputation. The first representation is by Henry Brougham, a former student at Edinburgh, lawyer, political reformer, man of science, and latterly Lord Chancellor of the United Kingdom, the chief legal officer of the realm; he was recalling the privilege he had felt of "being present while the first philosopher of his age was the historian of his own discoveries" (Brougham 1845: 348). The second representation, by John Kay, a noted and witty caricaturist of well-known Scots of the day, records features of the same annual lecture from Black's course, identifiable from the various apparatus on the lecture table. The third is by John Robison, professor of natural philosophy at Edinburgh, long a friend and colleague of Black, and now, after Black's death, in the toils of editing his friend's chemistry lectures for publication, his frustrations breaking out as he unburdened himself to James Watt. "[Dr. Black's] theory of Lime is tedious beyond bearing and the reader (of any information) cannot but see the keeping up of the great discovery till the very last" (Musson and Robinson 1969: 344).

Brougham's is a verbal, public representation, from a chapter in a book about notable intellects of the Georgian era. He sought to rescue Black's reputation from the unjust neglect and diminished luster which he perceived in the writings of chemists, particularly Louis-Bernard Guyton de Morveau. Brougham drew on his memories of Black, writing at a distance of half a century. He recalled Black's

FIGURE 8.1 Joseph Black lecturing at Glasgow. Etching by John Kay, 1787. Photograph by Universal History Archive/Getty Images.

remarkably dextrous skills of experimental demonstration at an advanced age, exhibiting difficult, dangerous, and delicate maneuvers while simultaneously explaining to the class just why the experiment had to be done this particular way. He further described Black's low-pitched yet distinct voice, elegance of diction, and the refined composure of his lecturing persona, altogether an encomium, highly effective in its own terms, delivered with the authority of a known public figure, a former Lord Chancellor.

From such representations the historian may gather valuable evidence unavailable elsewhere. Perhaps the most interesting of these, despite the fascination of the exhibition of experimental skill, is the way in which Brougham's reminiscence acknowledged that at such a late stage in his career Black was not merely a practised performer, but that this performance, precisely as in a theater, was a mode of self-representation which exhibited Black as an historical actor, re-enacting the discovery which had made his reputation forty years earlier – perhaps, Brougham noted, with "the very instruments he had then used" (Brougham 1845: 347–8). It was thus a specifically historical performativity that Black had adopted by the late 1780s, and this in turn makes a general point. Representations of chemistry could originate with chemists themselves, in their forms of self-representation: the ways in which they defined their science, expressed and exhibited its content, and located its practice and its value within particular historical and cultural settings.

John Kay's 1787 etching of the elderly Black also illustrates the fixed air lecture, primarily through the apparatus on the bench in front of Black, including the caged bird which will demonstrate the toxic effects of breathing fixed air, and the candle which will be extinguished by it. Black himself is lightly caricatured, the satire as affectionate as it is critical. The most notable feature is his smile, giving an air of self-satisfaction as he performs his polished re-enactment. This image can thus be historically aligned with Brougham's verbal account, but the tone differs substantially, its mild satire containing a psychological edge markedly absent from Brougham's encomium.

John Robison faced serious problems when, after Black's death in 1799, he undertook the editing of Black's lectures for publication (Black 1803). He grappled with the technical issue of rendering relevant an old-fashioned text, superseded by Antoine Lavoisier's chemical textbook, the *Traité élémentaire de chimie* (Lavoisier 1789), available to British readers in Robert Kerr's translation of 1790, a text whose "synthetic" presentation of chemistry aroused Robison's disapproval. Additionally, Black's own lecture notes were not in any kind of publishable state, many consisting of commentary on the numerous experimental demonstrations presented by Black to his students. Robison reached a low opinion of them – popular demonstrations, neither adapting to nor criticizing Lavoisian chemistry in any philosophically sufficient or coherently organized manner. The lecture on caustic alkali and fixed air caused him particular dismay.

What had worked for Brougham as a climactic lecture appeared simply embarrassing considered within the series of lectures: old news, redolent of another time. Little or none of this appeared in the eventually published two-volume *Lectures on the Elements of Chemistry, by Joseph Black, M.D.* (Black 1803). This presented Black as a cautious, analytical inductive philosopher, at times expressing methodological and critical views explicitly reminiscent of Robison's anti-Lavoisian sentiments as seen in his editorial correspondence

(Christie 1982: 47–62). This public act of reputational rescue performed by Robison's edition was, therefore, in representational terms, utterly at variance with the representations of Black in Robison's private correspondence. The evidence and opinions in the latter are a valuable corrective source for the historian of chemistry, but they also make a starkly unavoidable point for representational history, concerning just how and why public representations are produced at specific moments of historical time, the problems and contradictions negotiated in their formulation, and just what kinds and degrees of veracity and evidential reliability they may be accorded.

SELF-REPRESENTATION

Chemists in the eighteenth century faced several issues when they had occasion to speak or write of the nature of their science and its practice, in formal orations, introductory lectures or chemical texts. Prominent among these was chemistry's relation to alchemy, but also problematic in certain circumstances were its associations with artisanal manufacture, and its status as an authentic and independent science. In a period of increasingly assertive rationality, the association with alchemy's unfulfillable and delusive quest for the transmutational powers of the philosophers' stone, and its elements of mysticism and of fraudulence, could be embarrassing. Condemnation of this alchemical ancestry and consequent vindication of chemistry's empirical and rational commitments was thus not uncommon, but less predictable and of more representational interest were the efforts of some academic chemists to provide a more rigorous and nuanced account of chemistry's history.

Such a vindication could be found in the introductory history which Herman Boerhaave delivered to his chemistry students at the University of Leiden in Holland, which, while intensely critical of older alchemy, nonetheless gave a surprisingly positive account of the more modern figure Theophrastus von Hohenheim (Paracelsus). While condemning unsustainable claims of Paracelsian alchemical medicine, Boerhaave made a reasoned apologia for Paracelsus' skills and contributions to chemistry (Boerhaave 1727: 21–9). In the next generation, William Cullen also introduced his lecture courses at Glasgow and Edinburgh with a substantial history of the subject (Cullen n.d.: 1–30). It made a strongly argued case for the assignable origins of chemistry, not in the diverse crafts of metallurgy, pharmacy, and alchemy as Boerhaave claimed, but in the substantial and systematic pharmaceutical art of Arabian culture.

Cullen also condemned the alchemists, but equally commended aspects of Paracelsus' revolutionary career, welcoming its critical impact upon the orthodoxies of the medical tradition derived from the ancient Greco-Roman physician Galen, and the successes of Paracelsianism in establishing chemical medicine. He then advanced to the emergence of modern chemistry in the seventeenth century, emphasizing chemistry's progression to scientific status by way of acquiring rationally coherent theoretical constructs, as found in the corpuscular mechanism of Robert Boyle, and the force-inflected speculations of Isaac Newton (Christie 1994: 16–18). Much later John Robison would characterize such features of Cullen's own achievement, with respect to chemistry's status as science, appropriating it from the craft locations of "artists, metallurgists and pharmacists" and rendering it "a liberal science fit for gentlemen to study," a simultaneous transformation of its social and cognitive standing (Black 1803: xxii).

However, for Cullen, and for many other eighteenth-century chemists, this kind of transformation did not signify any severing of relations with the chemically based practical arts; rather, the relations were reformulated.[1] No eighteenth-century chemist argued for chemistry's irrelevance to the development of manufacturing and agricultural practice. Instead, the emphasis was upon the promise of enhanced economic utility offered by its analytical experimentation, systematic classification and organization of substances and their properties, and its theoretical knowledge of chemical causation. Such advocacy was variously expressed and institutionalized in the writings and activities of chemists, and can be found thoroughly exemplified in Germany, Britain, and France, the latter instanced by chemists such Pierre-Joseph Macquer's work at the Sèvres Porcelain Manufactory, Lavoisier's work at the Salpêtrière (government saltpeter manufactory), and Jean-Antoine Chaptal's founding of the Société pour l'Encouragement de l'Industrie Nationale in 1801.[2]

The utilitarian orientation of chemistry as a science was not merely incidental or opportunistic. It became a consistent and defining feature of the promotion of chemical science's public image. We may counterpose to it an image produced by a non-chemist of the artisanal association of chemical production. In an epistemic and linguistic register, Adam Smith emphasized the esoteric language of chemical labor:

> Why has the chemical philosophy in all ages crept along in obscurity ... while other systems, less useful and not more agreeable to experience, have possessed universal admiration for whole centuries together? ... Salts, sulphurs and mercuries, acids and alkalis, are principles which can smooth things to those only who live about the furnace; but whose most common operations seem, to the bulk of mankind, as disjointed as any two events which the chemists would connect together by them. These artists, however, naturally explained things by principles that were familiar to them.
> (Smith 1795: 21–2)

Smith's point is that the restricted technical language of the laboratory or workshop produces an impenetrable vocabulary which has no clarifying or explanatory value for anyone outside the artisanal practice itself; hence, although a useful practice, it remains disregarded because it provides no comprehensible image of chemical processes for anyone outside the practice.

The whole tone of Smith's passage, with its "artists," "furnace" and Paracelsian principles, "Salts, sulphurs and mercuries," implies association of chemistry with artisanal practice, and its consequence of incommunicability. His point has recognizable kinship with the comments of Gabriel-François Venel early in his article on chemistry in Denis Diderot's *Encyclopédie* (Venel 1753: 48). Such images were just what Smith's university colleague William Cullen was contemporaneously endeavoring to overturn. For Cullen, chemistry had become "the assay master to the arts in general," thoroughly integrated within a modern commercial economy (Cullen 1747). Venel's initial negativity and Smith's interjection, however, make the point that the general and harmonious chorus of modernizing systematics, progressive utility, and social relocation could nonetheless contain some cognitively dissonant notes.

LABORATORY REPRESENTATIONS

Smith's verbal representation also introduces the topic of the spaces of chemical practice. Chief among these is the laboratory, whose kinds and functions were various. There were manufacturing laboratories, such as Ambrose Godfrey's in Covent Garden,[3] and the London Apothecary Guild's large laboratory at Blackfriars. Universities such as Glasgow, Edinburgh, and Montpellier and public lecturers in Paris and London had workplaces essential for preparing experimental apparatus and materials for lecturing, and these could also function for the research purposes of the lecturers. Pharmacists' shops also required laboratory space for the chemical compounding of medicines. Some of pharmaceutical chemistry's esoteric procedures were targeted by Robert Dossie, a well-known chemist of the middle decades of the century who wrote *The Elaboratory Laid Open* (1758), significant for its critical orientation toward aspects of pharmaceutical manufacture and trade.

William Lewis produced a long serial work, the *Commercium Philosophico-Technicum* (1763), strongly featuring chemical dimensions of manufacturing production. This book contained an illustration of an imposing laboratory, considerable in size and furnished with a crowded and impressive array of contemporary instrumentation. Crowded as the laboratory was, it was nonetheless unpopulated by chemists. A glass worker is shown laboring beside a kiln, but only as part of an elevated image affixed to the rear-right wall. The large barometer to the right of the kiln image, the elevated placement of the image, and the rest of the instrumentation, including flaming but untended furnaces, all take on a slightly surreal and disorienting air, possibly originating in the perspectival dislocation induced by the visual depth of the kiln image within

FIGURE 8.2 Laboratory of William Lewis, *Commercium Philosophico-Technicum*, 1763. Sourced from Wikimedia Commons, Public Domain.

the laboratory image, such that doubts may arise as to this representation's full documentary realism as depiction of an actual, as opposed to a partly fictive, but not thereby falsified laboratory. It would be relevant to know whether Lewis had such a large image on his laboratory wall, what its source was, and why it was there. Its content, of active manufacturing production, reinforces the sense provided generally by the kinds of instrument representation around the laboratory, relatable to practically productive aspects of mid-eighteenth-century chemistry and mechanics.

The depopulated feature is repeated in an illustration of Joseph Priestley's laboratory, which possesses no personnel at all (it is available online, at knarf.english.upenn.edu/Gifs/prlab1.html). The reason these images appear depopulated has to do with historical comparison. Paintings of seventeenth-century alchemical laboratories tend to be populated not only by the alchemist, but by others – assistants or apprentices, onlookers, or perhaps the alchemist's wife or children. The chaotic interiors contain human action. These are humorous, at times satirical, representations, not formal illustrations of laboratories, but even the latter also contain personnel. As we move into the eighteenth century, the transition is to a large, well-ordered space, for instance with the image of Ambrose Godfrey's manufacturing laboratory in London (Morris 2015: 40–1). An illustration of the Dutch chemist and physician Conrad Barchusen's laboratory at Utrecht has the solitary Barchusen front center at a table, with chemical balance and weights (Morris 2015: 52).

FIGURE 8.3 Laboratory of Conrad Barchusen, der Sonnenborgh, Utrecht, from Barchusen, *Elementa chemiae* (Lugduni Batavorum, 1718). Courtesy of Science History Institute.

By the time we reach the Lewis and Priestley illustrations, human absence has become notable, and it is not immediately clear why this should be the case. One suggestion worth pursuing is a consideration of the way in which instrumentation appears to have become the increasingly privileged signifier of chemistry, a growing power and number of instruments now being used to stand for the character of the science (this is technically describable as a metonymic or synecdochic representational trope). In the eighteenth century, however, the story of laboratory illustration did not finally remain locked in such solely instrumental representation. Its humanity was rescued by the sketches of Marie Lavoisier, who spectacularly revived the populated laboratory and active experimental process in her acute sketches, which portray Petit Arsenal laboratory spaces, with individually identifiable experimenters performing precisely identifiable experiments (Beretta 2001: 43–52). Further aspects of Marie Lavoisier's representational work are considered below in the section on "Artistic Representation."

ALCHEMICAL PERSISTENCE

A notable example of alchemy's representational persistence is found in a painting of 1771 by Joseph Wright of Derby, *The Alchymist, in Search of the Philosopher's Stone, Discovers Phosphorus, and prays for the successful*

Conclusion of his Operation, as was the custom of the Ancient Chymical Astrology. For Wright, a well-known painter of scientific scenes such as *The Orrery* and *Experiment with a Bird and an Air Pump*, this was an unusually long-winded title. The latter paintings depicted modern eighteenth-century scenes of scientific demonstrations in domestic settings, with a particular esthetic focus on light, its sources, illuminative diffusion, occultation, and reflection. This compositional focus was a leading feature of his work, and it was continued in *The Alchymist*, a painting which indicates a turn to Gothic settings and subjects, at that time gaining cultural popularity. Did Wright have an impulse to represent a crucial but adventitious chemical moment of the discovery of phosphorus by Hennig Brandt, an "alchymist"? If so, what was the relation to the rational, industrializing progressivists of the Lunar Society acquainted with Wright, such as natural philosopher Erasmus Darwin and chemical manufacturer James Keir? Is the intentionality such as to isolate a moment of accidental discovery, itself a moment (as the title of the painting makes clear) in the traditional quest for the Philosophers' Stone? Is the glowing phosphoric light delusional, an *ignis fatuus*, with respect to the larger alchemical quest, and alchemy itself?

Part of the difficulty proceeding with such issues pertains to the painting's punctuality: all is present at once to the viewer's eye. It is informative then to think that Wright gave his painting a highly explicit, narrative temporal extension. The discovery of phosphorus is, by the painting's title, a stage of the quest for the Philosophers' Stone, for whose eventual fulfillment the alchymist now prays, in continuity with the ancient astrological alchemists, a continuity reflected in the explicit temporality represented by the clock and the moon, as the timing of the work was crucial in astrological alchemy. Because the alchymist prays, religious elements now inevitably intrude, reinforced pictorially by the alchymist's saint-like depiction and chapel-like laboratory. We thus have a painting of Gothic inclination, religiously inflected, of a moment of chemical discovery, and it is this discovery, the glowing phosphorus, which is the painting's primary illumination point, and through whose light the painting is composed.

To sort out paintings' meanings, both to painters and to potential viewers, is therefore a complex matter, and involves undecidable features, given the absence of direct and unequivocal testament. From the point of view of history of chemistry, however, it is not equivocal that alchemy still held, at the height of Enlightenment's scientific rationalism, an esthetic, cultural presence, however ambivalent that presence was, and that Gothic revivalism, explicitly present in the subject matter and the architectural detail of the painting, prolonged this alchemical presence in the work of an artist who also painted the popularization and domestication of modern astronomy and pneumatics. Further exploitation of the verbal and visual resources offered by alchemy are noted in later sections.

FIGURE 8.4 *The Alchemist in Search of the Philosophers Stone,* Joseph Wright of Derby, 1771. Derby Museum. Photograph by Christophel Fine Art/Universal Images Group via Getty Images.

THE PORTRAITURE OF CHEMICAL REVOLUTION

The esthetic culmination of the depiction of chemists in the eighteenth century is Jacques-Louis David's 1788 studio oil portrait of Antoine and Marie Lavoisier. This depiction catches the Ancien Régime chemical culture of France at its precipitous height immediately before the plunge into political revolution, during which Antoine will lose his life for deriving income from tax farming, but which Marie will survive, as will David, who would become the pre-eminent painter of the Revolutionary and Napoleonic periods. This painting also has its representational complexities, in its historical precision and detail, and its compositional form, which take it beyond the simple description of a conventional late Ancien Régime portrait of a scientifically and culturally prominent couple.

As Marco Beretta has shown with reference to the portrait, we have an historically specific moment, during the composition of Lavoisier's *Traité élémentaire*, the textbook of chemistry which incorporated his own research and that of colleagues on oxygen chemistry and the chemistry of gases more generally, reformed chemical nomenclature together with concepts of practical chemical analysis and chemical composition, and synthesized these fundamental changes in a new and teachable textbook (Lavoisier 1789; Beretta 2001: 27). This textbook came with thorough illustrations of the instrumentation of Lavoisier's laboratory, the illustrations provided by Marie Lavoisier.

Marie Lavoisier's artistic abilities were in part owing to the instruction in art she had undertaken with Jacques-Louis David. In addition to assisting in the laboratory, she had also already aided the French chemical campaign with widely admired translations of and commentary on recent English chemistry defending the phlogistic theory of combustion and calcination to which Lavoisier's oxygen chemistry was definitively opposed. Beretta successfully shows that Marie Lavoisier was no mere assistant who could draw appropriate instruments, but was a substantial participant in the campaign to establish the legitimacy of Lavoisian chemistry. The chemical instrumentation depicted, Beretta also argues, is plausibly seen as having a chronological arrangement, from the aereometer and its neighboring flask at the foot of the table, up to the gasometry instrumentation on the table, aligning the sequence of Lavoisier's research work from the early 1770s onward (Beretta 2001: 28–40).

David's portrait is thus more than a conventional work depicting a companionate, affective marriage from the philosophical *haute bourgeoisie* of Paris. The marriage is also shown as specifically collaborative. Lavoisier turns to gaze at his wife, whose portfolio, at the rear left of the picture, likely contains her recent illustrations. These depict the latest phase of scientific developments, whose represented instrumental basis in the portrait are the material artifacts which project an historical narrative of Lavoisier's research. Marie Lavoisier's ability as an illustrator, and by implication David's own teacherly excellence, is directly alluded to by the inclusion of Marie Lavoisier's portfolio, by which, and only by which, the visible designation of the instrumentation used for and appropriate to the new chemistry will reach a reading public.

FIGURE 8.5 Portrait of the French chemist Antoine Laurent de Lavoisier with his wife (Marie-Anne-Pierrette Paulze). Painting by Jacques Louis David, 1788. The Metropolitan Museum of Art, New York. Photograph by Leemage/Corbis via Getty Images.

Further analysis of the portrait's compositional features may help to supplement the foregoing account. The painting has unavoidably forceful compositional aspects. Obvious among them is the insistent diagonality focused on Antoine Lavoisier's extended, elegant black silk-stockinged right leg and the matching crease in the tablecloth, extended in the positioning of Marie Lavoisier's right arm, which, following the implications of Beretta's account of the historicity of the instrumentation, now includes not only Marie Lavoisier herself, but also her work.

The prone position of the flask on the floor allows David to maximize the virtuosic painting of the studio windows' reflection on the surface of the otherwise transparent glass of the instrument, and which thus brings glass, of different kinds according to the instrument they compose, and with respect to common and simultaneous optical properties of reflection and transparency, into closer esthetic focus. It may conceivably function as an additional lesson for Marie Lavoisier's illustrative ambitions: this is how you do it, or at least how it can be done with oil paint and canvas, as opposed to her less-versatile materials of production. The diagonality of leg, arm, and crease is taken up by the parallel diagonal line of direction of Lavoisier's eyes as he gazes at his wife. The diagonals are reinforced by the white quill in Lavoisier's hand, so we are also insistently at a moment of inscription, of composition of the text of which his wife's illustrations are a necessary component. At the rear left, on a chair, the edge of her portfolio again reinforces the parallel diagonals.

Antoine Lavoisier's gaze is the painting's primary focus, the vanishing perspectival point being somewhere between their heads, and thus the viewer intercepts the gaze, and the relation it may posit. This relation is vividly present, if not definitively clear. Lavoisier's gaze is interested, perhaps with a query, along with approval and/or gratitude. It is, however, unanswered by any return of gaze by Marie Lavoisier, who gazes indeed, but directly to the viewer, and necessarily to the painter (Vidal 1995: 620). She further occupies significantly more pictorial space than Lavoisier; moreover, posing on the window side of the studio, she receives more light and has more height than her seated husband, the contrast emphatically reinforced by the white and blue coloring of the left, feminine side of the painting, set off against the deep red and black coloring of the masculine side. The effect of height, lighting, and color contrast gives Marie Lavoisier, in her fetchingly informal dress, more presence than Lavoisier himself. The soft curves and folds of the delicately rendered dress compositionally mitigate the strong angularities of the masculine side of the painting, and, along with Marie Lavoisier's delicately pale skin tones, also serve as a decorous eroticization of the female figure.[4]

Further scrutiny of the directionality of the subjects' gazes provokes other questions. Lavoisier's gaze is reactive, to his wife. She, arm and hand with casual intimacy upon his shoulder, nonetheless does not await his response, but is more formally posed, face directly forward to the artist, and explicitly conscious of being portrayed. Lavoisier's gaze is part of the portrait's naturalistic depiction of a moment's action, within the picture and its narrative. Marie Lavoisier consciously gazes out of it, aware (she could not be otherwise) of the painterly processes which will produce her image, by one who has taught her how to produce images. Some of the complexities of David's painting here adumbrated may be arguable this way or that, but it is just this range and depth of historical, narrative reference, confidently and reflexively poised esthetic design, its adroit and precisely gendered rendition of the complex marriage of art with science, which ensure the portrait's status as a site of continuing relevance and interest for the history of chemistry's artistic representations.

A final, gender-related point occurs upon examining features of Marie Lavoisier's sketch of a respiration experiment of 1790–1791. This sketch captures a specific, active moment in the progress of a complex experiment. Marie Lavoisier depicts her own role in the laboratory, recorder of the experimental process, seated at the table, taking notes of what occurs. In other words, within her sketch that visually represents the experiment she represents herself in the act of producing the experiment's first and crucial verbal representation. Like her husband in David's portrait, she is posed at a table angled to her left, body facing the table; like him, she is gazing contra-posture, from left to right; and it is she who now possesses the act of writing. Moreover, the precise posing of her left arm, wrist, and hand closely mimics David's depiction of Lavoisier's left arm and exquisitely turned wrist. It thus becomes plausible to think of Mme. Lavoisier's sketch as being artfully in visual conversation with David's portrait; witty, lightly parodic, and pointedly ironic – who now has both the sketching pencil and the inscribing pen?

FIGURE 8.6 Drawing by Marie Lavoisier of an experiment on respiration, Salpetrière laboratory, 1790–1791. Science History Images/Alamy Stock Photo.

CHEMICAL SATIRE

There was a healthy relationship between satire and chemistry during the long eighteenth century, but the British scene of the 1790s serves as a particularly good case study. A significant feature of the plethora of political caricatures which characterized the febrile British politics of response to the revolution in France was the satirical use of materials drawn from chemical, even alchemical culture. The latter was used in James Gillray's figuration of first minister William Pitt's dissolution of parliament in 1796, *The Dissolution, or the Alchymist producing an Aetherial Representation* (available online at https://collections.nlm.nih.gov/catalog/nlm:nlmuid-101393181-img). Pitt the alchemist uses a bellows shaped as the monarchical crown to encourage the furnace which heats the flask in which parliament is literally dissolved and then vaporized. From the laboratory ceiling hang animal specimens typical of earlier images of alchemical laboratories, and the "Aetherial Representation" is a further image of Pitt himself, now sitting solitary and enthroned amid the vapors issuing from the flask. This is the visual reinforcement of the pun: Gillray represents Pitt etherializing parliamentary representation.

The counter-revolutionary tract *Reflections on the French Revolution* (1790) by Edmund Burke is notable for its rhetorical adaptation of chemical terms for politico-polemical purposes, a tactic which critically focused the radicalism of Joseph Priestley (Crosland 1987). This rhetoric was adumbrated and then continued in visual satire. Priestley appeared up to a dozen times in the caricatures of the period, in the first instance because of his publicly expressed political and religious views, which included vindication of the execution of Charles I during the mid-seventeenth century English Revolution, support for the American cause prior to and during the War of Independence, support for the revolution in France, and unceasing argument for abolition of the parliamentary Acts which excluded most religious Dissenters from civil office, as well as from graduation at Oxford or Cambridge. By the 1780s Priestley had become a leading public spokesman for the Dissenting cause, provoking the hostility of governments and of conservative opinion more generally within political culture. He was as well known for his early chemical researches on gases, successfully producing and identifying several new species, including "dephlogisticated air," renamed "oxygen" by Lavoisier. By the 1790s his scientific reputation had become closely attached to his long-term defense of phlogiston, the material agent of combustion and calcination, against whose existence Lavoisier's new oxygen chemistry was formulated.

The most effective visual satires of Priestley produced by the caricaturists contrived to amalgamate the chemical, political, and religious elements of his public persona. *Dr Phlogiston, the Priestley Politician or Political Priest* added to its punning title the image of an incendiary Priestley clutching in his raised hand some "Political Sermons," in the other hand his own *Essay on the first Principles of Government* (1768), both in a state of combustion. Beneath his feet is another work entitled "The Bible Explained Away," in his pocket his own work on matter and spirit, and another on gunpowder, referring to the designation Gunpowder Joe, a sobriquet he suffered from after writing an ill-advised polemic, referring to politically argumentative "grains of gunpowder" in Britain (Priestley 1785: 40–1).

Priestley suffered further at the hands of the leading political caricaturists of the day, James Gillray, Andrew Cruickshank, and James Sayers. In 1789 the latter produced an unusually complex image. While other caricaturists could tend to demonize Priestley's image, Sayers's *A Vision. The Repeal of the Test Act* (1789), produced in anticipation of the parliamentary debate of 1790 on the Act, has a more historically inflected and significantly varied set of characters. The three central figures in the church pulpit are the leaders of Rational Dissent: Priestley, Richard Price, and Theophilus Lindsey. In the congregation are, to the left, Charles James Fox, reforming Whig politician and principal parliamentary supporter of the motion to repeal the Act, seated center is Tom Paine, and to the right stands Lord Stanhope, sympathetic to the French revolutionaries. Stanhope's early political career had been supported by the Earl of Shelburne, also Priestley's aristocratic patron, not pictured but symbolically represented by the Shelburne family crest above the pulpit.

The seated figure toward bottom right is Richard Watson, author of the theologically liberal *Letters of a Christian Whig* (1772), formerly Regius Professor of Divinity at Cambridge, now the suspect Bishop of Llandaff. Lower left is a Jew, looking to despoil Church regalia. To the pulpit's right is a doorway to the Sancta Sanctorum, which contains a portrait of Oliver Cromwell. Sayers thus invokes England's revolutionary past in the genealogy of the contemporary cast of characters. The Jew's presence may indicate a reference to Priestley's recent attempts to persuade London Jewry toward the anti-Trinitarian monotheism common to both him and Lindsey. The latter is tearing up the Thirty-Nine Articles of the Church of England, by whose official governmentally supported presence at the heart of Anglicanism Dissenters suffered their civil exclusions owing to their principled refusal to subscribe to the Articles. Price bids the congregation to prayer in support of the efforts of the French National Convention. Priestley's fiery, smoky, phlogistic exhalation, labeled with "Socinianism," "Atheism," and "Materialism," threatens the angel flying past the window, behind whom a church steeple is being pulled down. Collectively these all form a complex image of the contemporary entanglement of religion and politics and its historical lineage. Priestley's phlogistonism is again his chief visual marker, pictorially expressive of his heterodox religious opinions.

FIGURE 8.7 James Sayers, "*A Vision. The Repeal of the Test Act*," 1790. Courtesy Alamy Stock Photo.

An unremarked feature of Sayers' cast of political and religious radicals is that no fewer than four of them, Priestley, Watson, Price, and Stanhope, were men of science: Price was a mathematician, Stanhope was a writer on electricity and a practical inventor, and Watson's theological and ecclesiastical career was preceded by teaching and writing on chemistry (there are thus two chemists in this image). All four, moreover, were Fellows of Britain's pre-eminent scientific society, the Royal Society of London. This may not be simply coincidental. Whether Sayers included natural scientists as such as a deliberate part of his satirical target, or whether their presence was simply contingent upon current roles with respect to politics and religion, the image nevertheless registers a set of associations to be found within sectors of the scientific community as a whole, and as an historical entity.

Such associations would include the relation between religious heterodoxy and natural science dating from the mid-seventeenth century, then strongly exemplified by Isaac Newton and some of his close associates, continuing in the careers of Priestley, Price, and other Rational Dissenters. Priestley's radicalism was paralleled at the time in the political orientation of three well-known chemists, Thomas Cooper, James Keir F.R.S., and Thomas Beddoes. A final observation on the content of *A Vision*, which may again pick out something inadvertent, but equally may not, is the chemical accuracy with which Sayers drew Priestley's exhalation. In Priestley's account of the respiration cycle, the final exhalation stage is indeed evacuation of the body's surplus phlogiston; thus, it is possible to think of this defining feature of Priestley's public image as providing not a fanciful but a literal chemical support for the allegorized content of the exhalation.

LITERARY REPRESENTATION

Priestley and his family undoubtedly suffered at the hands of the dominant reactionary political culture, the attacks in popular public media being followed by the Birmingham riots which burnt down his house and laboratory. After a short spell at Hackney College and the Hackney Gravel Pit Meeting, he and his family emigrated to America. One of his correspondents when in America was the poet Anna Letitia Barbauld, in her own time a well-known and widely admired writer. Priestley and her father had taught at the Dissenting Academy in Warrington, and since her girlhood she had been a close friend of the Priestley family, Priestley's wife Mary becoming something of a surrogate mother to her. She was also an educator, and at times as forthright if not as continuously provocative as Priestley, in the expression of her own, comparable, political views.

Priestley's appearances in Barbauld's poetry received gentler and more positive handling than from the caricaturists of the 1790s. He appeared in seven of her poems, written with varying degrees of both seriousness and critical humor, aware of and concerned for his public persona and the kinds of opposition it aroused, aware too of the range and content of Priestley's interests and preoccupations, including his chemistry. Some of the poems have precise domestic settings and chemical reference, such as "The Mouse's Petition," written at a time when Priestley used mice as part of his experimentation, and represents the mouse's plea for freedom from captivity and its experimental fate (Barbauld 1994: 36).

The tone of Barbauld's poem is predominantly humorous, but its themes, of captivity and liberty, and of sentient life and mind, give it an underlying political and moral edge which draws together scientific and political elements of Priestley's life, such that the poem expresses rather more than simple domestic humor. She folded the poem between the wires of the mouse's cage, for Priestley to find the following morning. "A Character of Joseph Priestley" exhibits more serious intent, specifying Priestley's moral and intellectual attributes in more formal Augustan style, meter, and imagery (Barbauld 1994: 37). Even here, however, although Priestley is Truth's fearless champion, "piercing," "bold," an "ardent genius" in his pursuit of knowledge, there are hints of necessarily critical understanding. Priestley is also "eccentric" in his quests for scientific and religious truth, in other words off-center and prone to irregular motion, and this eccentricity is also for Barbauld associated with the restless and undisciplined attribute of genius (Barbauld 1825: 166–7). Thus to characterize Priestley as an eccentric genius was for Barbauld by no means a straightforward commendation of character, and introduces a critically evaluative dimension to interrupt the measured Augustan progress of the verse.

A particularly informative, cogently conceived poem of the early 1770s is "An Inventory of the Furniture in Dr. Priestley's Study," a telling description of what meets the eye in Priestley's principal workplace (Barbauld 1994: 38–9). As such it functions equally to inventory the furniture, and the cast of Priestley's mind. The opening references most likely allude to Priestley's historical and biographical charts: chronological maps depicting historical empires, and lifetimes of notable historical figures, designed to give an overview of the temporal extensions of both subjects (Priestley 1769). Both were well received, and were a significant factor in the formation of his early intellectual reputation. Elsewhere a group portrait of British monarchs "on a packthread swings," a lightly veiled reference to Priestley's less than reverential disposition toward monarchy. A selection from Priestley's library then follows: works of the Church Fathers, a theological preoccupation of Priestley's; the Roman writers Ovid and Juvenal; and legal texts, among them probably those of Blackstone, with whom he argued, and of Hugo Grotius, the great seventeenth-century Dutch philosophical jurist whose work Priestley valued for theological reasons.

The poem then moves to scientific apparatus, "A shelf of bottles, jar and phial," introducing the Leyden Phial, about which Priestley had written extensively in his *History of Electricity* (1767), again a popular and successful work. So significant is the Leyden instrument and experiment in the development of electrical science as recounted in Priestley's *History* that it can be considered as the primary agent of fundamental change in that narrative as a whole. The unmentioned figure in the immediate background of the poem at just this point was Benjamin Franklin, whose conceptualization of positive and negative electricity had successfully theorized the unexpected, puzzling, and dangerous effects of the phial's condensation of electrical charge. Franklin had become a close acquaintance of Priestley and was the dedicatee of his *New Chart of History* (1769), their relationship being both scientific and political. Franklin is by implication immediately introduced into the poem in the line "All filled with lightning keen and genuine," for Franklin's wider public recognition rested on his invention of the lightning conductor, the appropriate form of which, Franklin's pointed American version, or the blunt British version, was a scientifically and politically contentious matter. Barbauld's lines now take a complex, literary, and quasi-alchemical turn:

All filled with lightning keen and genuine,
And many a little imp he'll pen you in;
Which, like Le Sage's sprite, let out,
Among the neighbours makes a rout;
Brings the lightning on their houses,
And kills their geese, and frights their spouses.

The penning of a little imp in a phial evokes the image of a homunculus, of magical–alchemical notoriety. The reference to Le Sage is not to a near contemporary, the Genevan natural philosopher, but to René Le Sage, author of *Le Diable Boiteux* (1707, altered and improved 1725), which recounted the release of a diabolic spirit, Asmodeus, by an amorous university student who, fleeing pursuers, finds himself in an astrologer's garret. Asmodeus removes the roofs of neighboring houses, to reveal the hidden lives of their inhabitants. The astrologer's room is closely comparable to Wright's *Alchymist*, and to Priestley's study: "books and papers were heaped up in confusion on the table; a globe and mariner's compass occupied one side of the room, and on the other were ranged phials and quadrants; all of which made him conclude that he had found his way into the haunt of some astrologer" (Le Sage [1707] 1841: 1). We see once again the resource provided by astrological–alchemical imagery, in the same year as Wright's painting, but now in frankly comic mode, rather than incipiently Gothic. Instead of the expected direct reference to Franklin and the rationality of safety from lightning strike, the figuration inverts expectation by synthesizing recent scientific culture with a literary culture familiar with the Le Sage tale from an English theatrical adaptation of 1768.

Le Sage's instrumental focus continues with a sensitive political thermometer:

> ... by which
> He settles, to the nicest pitch,
> the just degrees of heat to raise,
> Sermons, or politics, or plays.

The next twenty-six lines elaborate on the matter and form of Priestley's writings, describing a chaotic desk reflective also of the range and pace of his output: printer's proofs; "Answer, remark, reply, rejoinder"; just-written books awaiting the printer; others partly composed. One might suspect and excuse a degree of hyperbole in Barbauld's enumeration, but it is nonetheless the case that Priestley would often have multiple works in process of composition, on religion, philosophy, history, chemistry, and even literary theory.

Much of Priestley's work had argumentative and polemical intent, and its often rapid pace of composition and publication was entirely deliberate. Priestley held the conviction that in rational argumentative occupation of the public sphere lay the key to progress on the issues which preoccupied him, be it phlogistic chemistry or the non-divinity of Christ. This was the reason he remained with the London printer and publisher Joseph Johnson, who had a sophisticated distribution network, was not afraid of controversy, and above all was reliable for getting Priestley's work quickly into public circulation. Barbauld's lines capture something of this constant pace and diversity of Priestley's literary production, and are always conscious of its provocatively argumentative nature.

The section ends with an elaborate figuration, comparing his works to Cadmus' sowing of the dragon's teeth, from which arise, feet first, fierce warriors who commence to fight: "And all, like controversial writing, / Were born with teeth, and sprung up fighting." This Ovidian figure is rendered more complex by playing on the subject of dentition. The writings not only spring from the combative ferocity imparted by the dragon's teeth, but already themselves have teeth, as is the case at birth with some animals, and occasionally humans. The writings are born, that is, both from and with teeth, immediately motivated and equipped for combat, a doubled dentition.

Although the "Inventory" treats of central and serious themes which characterize Priestley's developing career, its intimate domestic setting, its adroit play with contemporary and classical imagery, and above all its critically distanced and ironically humorous diction give it a light and personal tone. This is reinforced prosodically by the rhyming of the couplets, which in addition to monosyllabic rhymes employ a number of bisyllabic and trisyllabic rhymes (e.g. rejoinder/coined here, or olio/folio). Some are in themselves amusing in their disjunctive, far-fetched form. Their collective, calculated effect is to add a jaunty, rhythmic bounce, which tends to undercut any pretentiously serious or sententious dimensions to the poem.

Indeed, the surprising turn taken by the concluding four lines appears to emphasize the poem's lack of gravitas, its ephemerality:

> "But what is this," I hear you cry,
> "Which saucily provokes my eye?"
> A thing unknown, without a name,
> Born of the air and doomed to flame.

This is an arresting end-point, and for at least two reasons. First, it appears to be a highly reflexive intrusion of the text of the poem into the study, eliciting an alarmed query, voiced by Priestley, certainly echoed by the reader – "this" is the poem itself, left on the desk for Priestley to find, as she had comparably left "The Mouse's Petition." Second, the last line appears to be a directly chemical allusion to the contemporaneous study of airs and combustion.

However, some caution is necessary here. "Born of air and doomed to flame": was Barbauld comparing her poetic efforts to oxygenic combustion, her manuscript cast into the fire, thus vanishing? This immediate implication may nonetheless be mistaken, for the current dating of the poem by its expert editors is to 1771, too early for reference to Priestley's discovery of "dephlogisticated air" (oxygen) and its properties. The editors offer instead, and plausibly enough, Henry Cavendish's recent discovery of "inflammable air" (hydrogen) (Barbauld 1994: 248). It is, however, difficult to think of inflammable air, in terms of the sources of its chemical derivation, as "born of air," as opposed to simply *being* an air, whereas dephlogisticated air was precisely a constituent of atmospheric air. The editors concede that the dating of the "Inventory" to 1771 is not conclusive, so the possible reference to dephlogisticated air is not completely excluded.

What cannot be excluded is the understanding of combustion at that time common to Priestley and Cavendish, and others of the British chemical community, quite possibly also to Barbauld, namely that the primary chemical agent of combustion of any material substance, be it oxygen- or hydrogen-related, was phlogiston. In such terms, the poem's paper manuscript is thus doomed to a phlogistic end. Nor, finally, should the gendered comparability of Barbauld with Marie Lavoisier be ignored. However differently inflected by the nature, form, and content of their artistic expression, both female artists successfully contrived their reflexive, esthetic introjection into the male-dominated world of chemical science and its representation. To do so, each used an act of specifically chemical inscription.

A rather different kind of chemico-literary interaction is provided by a writer whose literary stature is comparable to David's in the visual arts. Johann Wolfgang von Goethe's novel *Die Wahlverwandtschaften*, first published in 1809, is a nineteenth-century work, but has a particular relevance for the history of eighteenth-century chemistry, because its title, *Elective Affinities* in English translation, refers to a distinctive and widespread feature of eighteenth-century chemistry.[5] Elective affinity, or attraction, was a concept which was held to account for the way in which chemical substances combine in what appears to be a preferential manner, such that, in the simple case of a two-component compound AB, B is displaced by a third substance, C, during a chemical reaction, with the formation of a new compound, AC. According to this concept, the affinity exhibited between A and C is stronger than that of A and B. The notion was extended to reactions between two compounds, AB and CD, in which B and C are exchanged, thus forming the new compounds AD and BC.

The most common mode of presenting reactions-as-affinities was tabular. Beginning in the second decade of the eighteenth century, affinity tables became common, increasing in size and range in response to chemical discovery and to developments in practical methods of chemical analysis and synthesis. Their basic form would place a substance at the head of a column, and below it a series of reactive substances ranked in descending order of degree of affinity for the heading substance. The columns of an affinity table thus represented a displacement series, any substance being able to displace any below it already in combination with the heading substance. The potential of such tables for coherently organizing and presenting large amounts of chemical data and the regularities apparently governing their reactive behavior was well appreciated, and they became a significant didactic tool in chemical education. The concept of affinity also provoked theoretical speculation as to the nature of the cause underlying affinity relations. Influential from the middle decades of the century onward were Newtonian notions of chemical "attraction," forces analogous to gravitational or electrical attraction, but operating at very small distances, a micro-corpuscular scale of action.

Goethe was also a man of science as well as letters, and indeed wrote his own works of natural science (Krätz 1999). It is thus hardly surprising that he exhibited familiarity with the concept of chemical affinity. His novel has received mixed critical reactions for the last two centuries, to such an extent that scholars now write books on the critical responses to the novel. This chapter's interest has a closer focus, namely upon the role and function of the concept of elective affinities in the conceiving and compositional execution of the novel. The story concerns four people on an isolated estate. Eduard and Charlotte are an aristocratic married couple, each having been previously married. The Captain, a close friend of Eduard, comes to visit, and they are joined later by Ottilie, Charlotte's young niece. The result of these visitations is a change of partners for Eduard and Charlotte, Eduard moving from Charlotte to Ottilie, and Charlotte and the Captain engaging each other's affections. The results of this complex affective exchange are unhappy, and by the finish of the novel it has taken on the proportions and properties of a tragedy.

The novel's use of affinity is most often described as "metaphorical," the reader apparently encouraged by the text, and by some critics, to grasp the relations between the characters as analogous to the affinity relations exhibited by chemical reactivity, the characters thus in the grip of natural, law-like forces, reacting in an experimental chemical retort, the enclosed environment of the estate. The characters come to understand and express their affective relations in explicit terms of elective affinities.

Chapter 4 of the novel is an extended account and discussion of affinity, Eduard and the Captain endeavoring to render it comprehensible to Charlotte, who exhibits a degree of skepticism concerning the semantics of election as applied to the interactions of inanimate matter, and advises caution and critical reflection with respect to the implicit analogies induced by such terms.

> [T]hese comparisons are pleasant and entertaining; and who is there who does not like to play with analogy? ... If he has been somewhat liberal with such words as Election and Elective affinities he will do well to turn back again to himself, and take the opportunity of considering carefully the value and meaning of such expressions.
>
> (Goethe [1809a] 1900: 58)

The Captain notes ways in which vocabularies of reaction have tended to use affective terms, even though this may not literally attribute animistic causality to matter, and proceeds to describe a double elective affinity reaction: "In this forsaking and flying we believe we are indeed observing the effects of some higher determination; we attribute a sort of will and choice to such creatures, and really feel justified in using technical words and speaking of 'Elective Affinities'" (59) He elaborates with a formulaic account of double affinities:

> I can put my meaning together with letters. Suppose an A so closely united with a B that all sorts of means, even violence, have been made to separate them, without effect. Then suppose a C in exactly the same position with respect to a D. Bring the two pairs into contact; A will immediately fling himself on D, C on B, without it being possible to say which had first left its first connection, or made the first move toward the second.

Eduard endeavors to explain further, personifying the initial case of single displacement:

> ... we will look upon the formula as an analogy, out of which we can devise a lesson for immediate use. You stand for A, Charlotte, and I am your B; really truly I cling to you, I depend on you, I follow you, just as B does A. C is obviously the Captain, who at present is in some degree withdrawing me from you.
>
> (60–1)

So that Charlotte will not be left in solitude, Eduard immediately suggests an additional formula term, D, to denote Ottilie, who has not yet arrived. This appears to transform the reaction formula to an incipient double displacement, four components united in pairs which upon reaction change their pairings, thus setting a structural template for the narrative that follows.

In more general terms, when Goethe wrote an anonymous "Advertisement" for the novel, it laid emphasis upon its chemical derivation:

> It appears that this odd title was suggested to the author by experiments he conducted in the physical sciences. Possibly he noticed that, in the natural sciences, ethical analogies are used to draw closer to the circle of human knowledge things far distant from it; and probably was all the more inclined, in an ethical instance, to refer a chemical discourse of analogies back to its spiritual origin, indeed because everywhere there is but One Nature, and the traces of disturbing passionate necessity run ceaselessly through the calm empire of rational freedom, traces that can be entirely extinguished only by a higher hand, and perhaps not in this life.
>
> (Goethe 1809b)

Clearly, the "Advertisement" focuses the content of discussion and argument of Chapter 4, concerning the questionable validity of chemical–ethical analogy, but adds a validating transcendental concept, "One Nature," implying some form of underlying unity, however hard to discern, between the physical and the ethical.

The novel's structural analogy might seem, therefore, to offer a ready key for understanding the compositional features of plot and character. Nonetheless, certain obvious difficulties may persist with any straightforward application, and these pertain to the scope and the detail of the analogy (Adler 1990: 268–75).[6] Are we to understand, for instance, the working out of one double affinity reaction, or of a sequence of single affinity reactions? Further, which particular version of affinity chemistry is in play? Goethe was undoubtedly familiar with the most influential treatment of the subject in the latter decades of the eighteenth century, namely Torbern Bergman's *Disquisitio de attractionibus electivis* (1775).

However, at the turn of the nineteenth century, Bergman's work came under the critical scrutiny of the prominent chemist Claude-Louis Berthollet, in his *Recherches sur les loix de l'affinité* published in 1801. Berthollet's critique did not advocate complete abandonment of the concept, but questioned its law-like status, as taken for granted by Bergman, and announced that his aim was to show "that elective affinity, in general, does not act as a determinate force" (Berthollet 1801: 5). To do this he examined the ways in which other regular features of reactivity, such as mass, temperature, and degree of saturation definitively affect the outcomes of reactions. This was a complex critical confrontation, in that Bergman was aware of such features, and that Berthollet was not intent upon wholesale critical annihilation of affinity.

For Berthollet, what was at issue was Bergman's assumption of affinities as invariant, asserting a uniform causal force which gave determinate necessity to the outcomes of reactions. Berthollet proposed that the specific actions of affinities were contingent upon the additional factors he adduced. In comparison with Bergman, Berthollet had changed the modal status of affinity relations, from invariant necessity to contingent variables. Given the accounts of and disposition towards affinity voiced in Chapter 4 of Goethe's novel, it is thus of immediate interpretive consequence whether Goethe had familiarity with and favorable inclination towards Berthollet's critique, and given Goethe's own scientific interests and credentials, it is by no means unlikely that he knew of it. If he did, then it casts the Captain's determinist account of affinity in a rather different light, as outmoded, even to some degree mistaken. The Captain actually admits his understanding of chemical affinity may be outdated and superseded. The chemically up-to-date reader would realize that reactive relations in chemistry are much more complex, and such a reader might wish to pursue the affective and ethical analogues of this more complex chemistry to reach a reformulated understanding of the novel.

Such extensions of the referential scope of analogical or "metaphorical" readings of the novel can serve to refocus the rhetorical function of affinity for the novel. Although metaphor is a figure which works by comparison, its encompassing function for the work as a whole resembles what is termed a "conceit," originally understood with reference to Renaissance literature as a rhetorical figure establishing a form capable of sustained expression of theme and content throughout a work. A conceit may be of a metaphorical or an allegorical type, whose ground of comparison may indeed be fanciful or far-fetched. It is not, however, necessarily a comparative figure, and only later does the term mutate to a critical connotation, a "mere conceit," perhaps exhibiting virtuosity or wit, but too fanciful or flimsy to attain esthetic or cognitive conviction.

In its pre-critical connotation, the literary deployment of affinity may qualify as a conceit, and Goethe's novel as a revival of the figure's rhetorical potential. Yet if it is, its functionality for the novel still requires further qualification. Chapter 4 is not primarily a set of notes to aid the reader, although it has that subsidiary effect, whatever ambivalences or misconceptions it contains. In the first instance, "affinity" is already a theoretical conceit in chemical science. Additionally, the conceit itself does not simply function as a conceptual template rhetorically imposed. It is instead internalized to become part of the action, a plot function, in just the sense that the chapter registers the characters' knowledge of and attitudes toward the affinity concept. It thus becomes part, misconceived or not, of their self-understanding, their collective relational comprehension, and some degree of justificatory conceptualization of motivations. Part of the way affinity acts in the novel is insofar as the characters come to conceive it as an active principle in human relations, and this psycho-activation, an internalized, reflexive use of the conceit, distinguishes the novel's rhetorical originality.

CONCLUSION

This chapter has endeavored to show that representations of chemistry and chemists in the eighteenth century were both various and complex, and to a degree unpredictable. They include attempts by chemists to promote positive public images of a progressive science; they range from canonical painting to popular caricature, from domestic poetry to canonical novel. They deploy satire and irony, allegory, metaphor, and metonymy as tropes of representation. They develop senses of chemistry's immediate and longer-term historicity as they become engaged in Gothic revivalism, in formal historiography, or in the contemporary history of the Chemical Revolution and the French Revolution. In so doing they could draw upon distinctive features of chemical practice and theory: lecture theater, laboratory, experiment and instrumentation, gas chemistry, phlogiston, and affinity. As the preceding chapters of this volume attest, all this occurs as chemistry expanded its conceptual and practical content, its institutional presence in universities, scientific academies, societies devoted to the improvement of agriculture and manufacture, governmental and quasi-governmental offices, and the publishing market. It was this growth in public presence and significant visibility which underlay the science's registration in practices of representation, as chemistry, inclusive of its alchemical past, came to serve a wide variety of expressive opportunities, each inflected by current contingencies of religion and politics, gender and social status, artistic and literary esthetics, within the eighteenth-century public sphere.

NOTES

CHAPTER 1

1   Stahl (1730) is an English translation by Peter Shaw of an early lecture by Stahl, based on a publication by one of his students.

CHAPTER 2

1   A harsher and influential attack on the analytical adequacy of distillation had already appeared in Boyle's *Sceptical Chymist* (1661), in which he argued that fire-based operations produced not elements but newly formed compounds, that the application of fire and strong heat destroyed the properties of the constituents beyond their meaningful empirical recognition and utility.
2   While explanations based on "principles" were also abstract, they were at least based on some qualitative correspondence based on known materials, e.g. "solid" salt, "active" mercury, or "inflammable" sulfur. Atoms and corpuscles were by definition imperceptible, although easily picturable, and possessed only primary qualities.
3   Determining the role of instruments in the history of science is a highly context-dependent exercise. Even seemingly simple instruments like the telescope and the thermometer were not passive, and their uses and applications were neither straightforward nor self-evident. The "proper functioning of an instrument and the interpretation of its readings are constructed through negotiation." Only when its "functional identity is established" does an instrument become "transparent," so to speak (Roberts 1991a: 200; Boantza 2013a).
4   It should be noted that "earthy substances" were still present in Lavoisier's "table of simple substances," which appeared in his 1789 *Elements of Chemistry* (A.-L. Lavoisier 1790: 176).

CHAPTER 3

1   The main discoveries were: cobalt (1735), platinum (1748), nickel (1751), bismuth (1753), magnesium (1754), hydrogen (1766), nitrogen (1772), barium

(1772), oxygen (1773–1774), chlorine and manganese (1774), methane (1777), molybdenum (1778), tungsten (1781), tellurium (1782), strontium (1787), uranium (1789), titanium (1791), yttrium (1794), chromium (1797), and beryllium (1798).

2 Johan Afzelius, "Förteningar på de till Chemiska Professiones in Uppsala hörande smalingar, hvilka Kongl. Academien, dels före, dels under Framl. Hr: Profess. Och Riddaren Bergmans tjenstetid inkjöpt dels också sedermera af Dess Enkefru blifvit till Kongl. Academien uplåtne," Uppsala University Library, Ms D 1469 fol. 4nn + 196 fols + 1 nn. The first 131 fols. contain a detailed inventory of Bergman's library (books, manuscripts, and maps and engravings). This is followed by the inventory of Bergman's laboratory, divided in several sections (fols. 133–196 + 1 nn).

CHAPTER 4

1 I would like to thank John Christie for his discussion and revision of this chapter.
2 Such indicators definitely refute the teleological accounts which reduce eighteenth-century chemistry to a pre-Lavoisieran period.
3 The term "revolution" nevertheless occurred twice in the entry "chemistry" of the *Encyclopédie*: as a return to a past situation, and as an aspiration for a radical change. "A revolution that would give Chemistry the rank it deserves, that would put it at least at the same rank as mathematical Physics, can be accomplished only by a courageous and enthusiastic chemist." This refers to a change in the public image of chemistry, rather than to a "paradigm shift," to use an anachronistic phrase.
4 The vision of nature as a laboratory was actually concretized when Italian chemists used the Vesuvius volcano as a natural laboratory to practice chemical investigations (Guerra 2015).
5 Kant adopted Stahl's theory of principles until 1787. When he updated his notions of chemistry in 1785 on the occasion of his physics lectures at the University, he became aware of Lavoisier's attacks against phlogiston. He adopted Lavoisier's theory of combustion in the 1790s.
6 Schelling criticized mechanical philosophy for reducing nature to mathematical abstractions and depriving matter of qualities and forces. The *Naturphilosoph* sought to comprehend nature as a unity endowed with immanent forces and qualities. *Naturphilosophie* required an empirical approach to nature, because sensory qualities are not reducible to mathematical understanding.

CHAPTER 8

1 For a wide-ranging analysis of British chemistry in the eighteenth-century public sphere, see Jan Golinski, *Science as Public Culture: Chemistry and Enlightenment in Britain, 1760–1820* (Cambridge: Cambridge University Press, 1992).
2 For the case of Germany, see Karl Hufbauer, *The Formation of the German Chemical Community, 1725–1790* (Berkeley: University of California Press, 1982), chapters 2–3.
3 For laboratory images, see Peter Morris, *The Matter Factory: A History of the Chemistry Laboratory* (London: Reaktion Books, 2015), 40–1. Chapters 1 and 2 contain many other relevant illustrations of early modern laboratories.
4 My analysis of David's depiction of Mme. Lavoisier is indebted to both Beretta and Vidal. However, I do not pursue the notions of Mme. Lavoisier as a mediating figure

between art and science, nor as symbolic of "Nature." I stress instead the reflexive relations inherent in the composition, and, given the exquisite artifice of Mme. Lavoisier's appearance, think it difficult to sustain this appearance as symbolic of "Nature."

5   For an English translation, I have used an old text (Goethe 1900) in preference to more recent translations, because of the relative clarity of the chemical affinity language in chapter 4, and its reference back to Bergman's ABCD notations for symbolic abstraction.

6   For further exploration of this point, useful treatment may be found in Jeremy Adler, "Goethe's Use of Chemical Theory in His *Elective Affinities*," in Andrew Cunningham and Nicholas Jardine (eds), *Romanticism and the Sciences* (Cambridge: Cambridge University Press, 1990), 263–79.

BIBLIOGRAPHY

Abbri, Ferdinando. 1991. *Science de l'air. Studi su Felice Fontana*. Cosenza: Brenner.
Abbri, Ferdinando, and Bernadette Bensaude-Vincent (eds). 1995. *Lavoisier in European Context: Negotiating a New Language for Chemistry*. Canton, MA: Science History Publications.
Abney Salomon, Charlotte A. 2019. "The Pocket Laboratory: The Blowpipe in Eighteenth-Century Swedish Chemistry." *Ambix*, 66: 1–22.
Adair, James M. 1790. *Essays on Fashionable Diseases*. London: Bateman.
Adler, Jeremy. 1990. "Goethe's use of Chemical Theory in his *Elective Affinities*." In Andrew Cunningham and Nicholas Jardine (eds), *Romanticism and the Sciences*. Cambridge: Cambridge University Press.
Airaksinen, Timo. 2010. "Active Principles and Trinities in Berkeley's Siris." *Revue philosophique de la France et de l'étranger*, 135: 57–70.
Albritton-Jonsson, Fredrik. 2013. *Enlightenment's Frontier: The Scottish Highlands and the Origins of Environmentalism*. New Haven, CT: Yale University Press.
Allen, Tim, Mike Cotterill, and Geoffrey Pike. 2001. "Copperas: An Account of the Whitstable Copperas Works and the First Major Chemical Industry in England." *Industrial Archaeology Review*, 23: 93–112.
Anderson, Robert G.W. 1978. *The Playfair Collection and the Teaching of Chemistry at the University of Edinburgh 1713–1858*. Edinburgh: Royal Scottish Museum.
Anderson, Robert G.W. 2006. "Boerhaave to Black: the Evolution of Chemistry Teaching." *Ambix*, 53: 237–54.
Anderson, Robert G.W. 2010. "Chemistry Beyond the Academy: Diversity in Scotland in the Early Nineteenth Century." *Ambix*, 57: 84–103.
Anderson, Robert G.W. (ed.). 2015. *Cradle of Chemistry. The Early Years of Chemistry at the University of Edinburgh*. Edinburgh: John Donald.
Anderson, Robert G.W. 2017. "Facts or Fantasies in the Chemistry Lecture Theatre?" In Jed Buchwald and Larry Stewart (eds), *The Romance of Science*. Cham: Springer.
Anderson, Robert G.W., and Jean Jones (eds). 2012. *The Correspondence of Joseph Black*. 2 vols. Farnham: Ashgate.

Anderson, Wilda. 1984. *Between the Library and the Laboratory: The Language of Chemistry in Eighteenth-Century France*. Baltimore, MD: Johns Hopkins University Press.

Barbauld, Anna Letitia. 1825. "The Hill of Science: A Vision." In Lucy Aikin (ed.), *The Works of Anna Letitia Barbauld*, vol. 2. London: Longman.

Barbauld, Anna Letitia. 1994. *The Poems of Anna Letitia Barbauld*. Edited by William McCarthy and Elizabet Kraft. Athens, GA: University of Georgia Press.

Barnett, Lydia. 2015. "The Theology of Climate Change: Sin as Agency in the Enlightenment's Anthropocene." *Environmental History*, 20: 217–37.

Barraclough, Kenneth C. 1991. "Steel in the Industrial Revolution." In Ronald F. Tylecote and Joan Day (eds), *The Industrial Revolution in Metals*. London: Institute of Metals.

Baumé, Antoine. 1773. *Chymie expérimentale et raisonnée*, 3 vols. Paris: Didot.

Baumé, Antoine. 1775. *Prix courants des préparations de chymie et de pharmacie qui se trouvent à Paris, chez Baumé, apothicaire, rue Coquillière*. Paris.

Belhoste, Bruno. 2011. *Paris savant: Parcours et rencontres au temps des Lumières*. Paris: Colin.

Bensaude-Vincent, Bernadette. 1990. "A View of the Chemical Revolution through Contemporary Textbooks: Lavoisier, Fourcroy and Chaptal." *British Journal for the History of Science*, 24: 435–60.

Bensaude-Vincent, Bernadette. 1992. "Between Chemistry and Politics: Lavoisier and the Balance". *The Eighteenth Century: Theory and Interpretation*, 33: 217–37.

Bensaude-Vincent, Bernadette. 1993. *Lavoisier: Mémoires d'une révolution*. Paris: Flammarion.

Bensaude-Vincent, Bernadette. 2010. "Lavoisier, lecteur de Condillac." *Dix-huitième siècle*, 42: 49–65.

Bensaude-Vincent, Bernadette. 2018. "Teaching Chemistry in the French Revolution: Pedagogy, Materials and Politics." In Lissa Roberts and Simon Werret (eds), *Compound Histories: Materials, Governance and Production, 1760–1840*. Leiden: Brill.

Bensaude-Vincent, Bernadette, and Ferdinando Abbri (eds). 1995. *Lavoisier in European Context. Negotiating a New Language for Chemistry*. Sagamore Beach, MA: Science History Publications.

Bensaude-Vincent, Bernadette, and Bruno Bernardi (eds). 2003. *Rousseau et les sciences*. Paris: Lharmattan.

Bensaude-Vincent, Bernadette, and Christine Blondel (eds). 2008. *Science and Spectacle in the European Enlightenment*. Aldershot: Ashgate.

Bensaude-Vincent, Bernadette, and Christine Lehman. 2007. "Public Lectures of Chemistry in Mid-Eighteenth-Century France." In Lawrence Principe (ed.), *New Narratives in Eighteenth-Century Chemistry*. Dordrecht: Springer.

Bensaude-Vincent, Bernadette, and Jonathan Simon. 2009. *Chemistry: The Impure Science*. London: Imperial College Press.

Bensaude-Vincent, Bernadette, and Isabelle Stengers. 1996. *A History of Chemistry*. Cambridge, MA: Harvard University Press.

Beretta, Marco. 1993. *The Enlightenment of Matter: The Definition of Chemistry from Agricola to Lavoisier*. Sagamore Beach, MA: Science History Publications.

Beretta, Marco. 2001. *Imaging a Career in Science: The Iconography of Antoine Laurent Lavoisier*. Canton MA: Science History Publications.

Beretta, Marco (ed.). 2005. *Lavoisier in Context*. Munich: Deutsches Museum.

Beretta, Marco. 2009. "Big Chemistry: Lavoisier's Design and Organisation of his Laboratories." In Marta Lourenço and Ana Carneiro (eds), *Spaces and Collections in the History of Science*. Lisbon: Museum of Science of the University of Lisbon.

Beretta, Marco. 2011. "Rinman, Diderot, and Lavoisier: New Evidence Regarding Guillayme François Rouelle's Private Laboratory and Chemistry Course." *Nuncius*, 26: 355–79.

Beretta, Marco. 2012a. "The Rise and Fall of the Glassmaker Paul Bosc d'Antic (1753–1784)." *Annals of Science*, 69: 375–93.

Beretta, Marco. 2012b. "Secrecy, Industry and Science: French Glassmaking in the Eighteenth Century." In Jed Buchwald (ed.), *A Master of Science History: Essays in Honor of Charles Coulston Gillispie*. Dordrecht: Springer.

Beretta, Marco. 2012c. "Imaging the Experiments on Respiration and Transpiration of Lavoisier and Séguin: Two Unknown Drawings by Madame Lavoisier." *Nuncius*, 27: 163–91.

Beretta, Marco. 2014a. "Unveiling Glass's Mysteries: Lavoisier, Loysel and the First Chemical Treatise on Glass (1765–1799)." In Ursula Klein and Carsten Reinhardt (eds), *Objects of Chemical Inquiry*. Sagamore Beach, MA: Science History Publications.

Beretta, Marco. 2014b. "Between the Workshop and the Laboratory: Lavoisier's Network of Instrument Makers." *Osiris*, 29: 197–214.

Beretta, Marco, and Paolo Brenni. 2022. *The Arsenal of Eighteenth Century Chemistry. The Laboratories of Antoine Laurent Lavoisier* (1743–1794). Leiden: Brill. forthcoming.

Berg, Maxine. 2004. "Consumption in Eighteenth- and Early Nineteenth-Century Britain." In Roderick Floud and Paul Johnson (eds), *The Cambridge Economic History of Modern Britain. Vol. 1, Industrialisation, 1700–1860*. Cambridge: Cambridge University Press.

Berg, Maxine. 2005. *Luxury and Pleasure in Eighteenth-Century Britain*. Oxford: Oxford University Press.

Berg, Maxine. 2010. "The British Product Revolution of the Eighteenth Century." In Jeff Horn, Leonard Rosenband, and Merritt Roe Smith (eds), *Reconceptualizing the Industrial Revolution*. Cambridge, MA: MIT Press.

Bergman, Torbern. 1775. "Disquisitio attractionibus electivis." *Nova Acta Regiae Societatis Scientiarum Upsaliensis*, 2: 161–270.

Bergman, Torbern. 1779. *Commentatio de tubo ferruminatorio, eiusdemque usu, in explorandis corporibus praesertim mineralibus*. Vienna: Ioann. Paul Krause.

Bergman, Torbern. 1782. *Sciagraphia regni mineralis: secundum principia proxima digesti*. Leipzig and Dresden: In biliopolio eruditorum.

Bergman, Torbern. 1965. *Torbern Bergman's Foreign Correspondence*. Edited by Göte Carlid and Johan Nordström. Stockholm: Almqvist & Wiksell.

Bergman, Torbern. 1985. "Autobiography." In J.A. Schufle, *Torbern Bergman: A Man Before his Time*. Lawrence, KS: Cornado.

Bernardi, Bruno. 2003. "La place des référents scientifiques dans l'invention conceptuelle: Une étude de cas." In Bernadette Bensaude-Vincent and Bruno Bernardi (eds), *Rousseau et les sciences*. Paris: L'Harmattan.

Berthelot, Marcellin. 1876. *La synthèse chimique*. Paris: Baillière.

Berthollet, Claude-Louis. 1801. *Recherches sur les lois de l'affinité*. Paris: Baudouin.

Bertomeu-Sánchez, José, and Antonio García-Belmar. 2000. "Spanish Chemistry Textbooks, 1788–1845." In Anders Lundgren and Bernadette Bensaude-Vincent

(eds), *Communicating Chemistry: Textbooks and Their Audiences, 1789–1939*. Sagamore Beach, MA: Science History Publications.
Bertucci, Paola. 2017. *Artisanal Enlightenment: Science and the Mechanical Arts in Old Regime France*. New Haven, CT: Yale University Press.
Black, Joseph. 1803. *Lectures on the Elements of Chemistry*. 2 vols. Edited by John Robison. London: Longman, and Edinburgh: Creech.
Black, Joseph. 1966. *Notes from Doctor Black's Lectures on Chemistry 1767/8*. Edited by Douglas McKie. Wilmslow: Imperial Chemical Industries.
Black, Joseph. 2012. *The Correspondence of Joseph Black*. Edited by Jean Jones and Robert G.W. Anderson. 2 vols. London: Routledge.
Boantza, Victor D. 2013a. *Matter and Method in the Long Chemical Revolution: Laws of Another Order*. Farnham: Ashgate.
Boantza, Victor D. 2013b. "The Rise and Fall of Nitrous Air Eudiometry: Enlightenment Ideals, Embodied Skills, and the Conflicts of Experimental Philosophy." *History of Science*, 5: 377–412.
Boantza, Victor D. 2017. "Elements, Instruments, and Menstruums: Boerhaave's Imponderable Fire Between Chemical Masterpiece and Physical Axiom." In Jed Buchwald and Larry Stewart (eds), *The Romance of Science: Essays in Honor of Trevor H. Levere*. Dordrecht: Springer.
Boerhaave, Herman. 1727. *A New Method of Chemistry: Including the History, Theory, and Practice of the Art*. London: Osborn and Longman.
Boerhaave, Herman. 1732. *Elementa Chemiae*. 2 vols. Leiden: Isaacum Severnium.
Boerhaave, Herman. 1735. *Elements of Chemistry*. London: Pemberton.
Boerhaave, Herman. 1754. *Eléments de chimie*. Paris: Chandon & fils.
Boklund, Uno. 1957. "A Lost Letter from Scheele to Lavoisier." *Lychnos*, 17: 1–24.
Bougard, Michael. 1999. *La chimie de Nicolas Lemery*. Turnhout: Brepols.
Bouvet, Maurice. 1937. *Histoire de la pharmacie en France*. Paris: Editions Occitania.
Bowler, Peter J. 2000. *The Earth Encompassed: A History of the Environmental Sciences*. London: Norton.
Boyle, Robert. 1661. *The Sceptical Chymist*. London: J. C.
Boyle, Robert. 1674. *Tracts: Containing Suspicions about Some Hidden Qualities of the Air*. London: W. G.
Bret, Patrice. 2016. "The Letter, the Dictionary and the Laboratory: Translating Chemistry and Mineralogy in Eighteenth-Century France." *Annals of Science*, 73: 122–42.
Brimblecombe, Peter. 2011. *The Big Smoke: A History of Air Pollution in London Since Medieval Times*. London: Routledge.
Brock, William H. 1993. *The Norton History of Chemistry*. New York, NY: Norton.
Brooke, John H., and Geoffrey N. Cantor. 2000. *Reconstructing Nature: The Engagement of Science and Religion*. Oxford: Oxford University Press.
Brougham, Henry. 1845. *Lives of Men of Letters who Flourished in the Age of George III*. London: Knight.
Buchanan, Brenda J. 2006. "Saltpetre: A Commodity of Empire." In Brenda J. Buchanan (ed.), *Gunpowder, Explosives and the State: A Technological History*. Aldershot: Ashgate.
Buffon, Georges-Louis Leclerc, Comte de. 1765. *Histoire naturelle générale et particulière, avec la description du Cabinet du Roi*, vol. 13. Paris: Imprimerie royale.
Burt, Roger. 1995. "The Transformation of the Non-Ferrous Metals Industries in the Seventeenth and Eighteenth Centuries." *The Economic History Review*, 48: 23–45.

Cavendish, Henry. 1766. "Three Papers, Containing Experiments on Factitious Air." *Philosophical Transactions of the Royal Society*, 56: 141–84.

Chalmers, Alan. 2012a. "Klein on the Origin of the Concept of Chemical Compound." *Foundations of Chemistry*, 14: 37–53.

Chalmers, Alan. 2012b. "Intermediate Causes and Explanation: The Key to Understanding the Scientific Revolution." *Studies in History and Philosophy of Science*, 43: 551–62.

Chandler, Anne. 2006. "Ann Radcliffe and Natural Theology." *Studies in the Novel*, 38: 133–53.

Chang, Hasok. 2012. *Is Water H2O? Evidence, Realism and Pluralism*. Heidelberg: Springer.

Chang, Hasok. 2015. "The Chemical Revolution Revisited." *Studies in History and Philosophy of Science*, 49: 91–98.

Chang, Ku-Ming. 2002. "Fermentation, Phlogiston and Matter Theory: Chemistry and Natural Philosophy in Georg Ernst Stahl's Zymotechnia Fundamentalis." *Early Science and Medicine*, 7: 31–64.

Christie, John R.R. 1981. "Ether and the Science of Chemistry, 1740–1790." In Geoffrey N. Cantor and Michael J.S. Hodge (eds), *Conceptions of Ether: Studies in the History of Ether Theories, 1740–1900*. Cambridge: Cambridge University Press.

Christie, John R.R. 1982. "Joseph Black and John Robison." In Allen D.C. Simpson (ed.), *Joseph Black, 1728–1799: A Commemorative Symposium*. Edinburgh: Royal Scottish Museum.

Christie, John R.R. 1994. "Historiography of Chemistry in the Eighteenth Century: Herman Boerhaave and William Cullen." *Ambix*, 41: 4–19.

Christie John R.R. 2015. "'The Most Perfect Liberty': Professors and Students in the Age of the Chemical Revolution." In Robert Anderson (ed.), *Cradle of Chemistry: The Early Years of Chemistry at the University of Edinburgh*. Edinburgh: John Donald.

Christie, John R.R. 2017. "Chemical Glasgow and its Entrepreneurs." In Lissa Roberts and Simon Werrett (eds), *Compound Histories: Materials, Governance and Production, 1760–1840*. Leiden: Brill.

Christie, John R.R., and Jan Golinski. 1982. "The Spreading of the Word: New Directions of the Historiography of Chemistry, 1600–1800." *History of Science*, 20: 235–66.

Clericuzio, Antonio. 2006. "Teaching Chemistry and Chemical Textbooks in France: From Beguin to Lemery." *Science and Education*, 15: 335–55.

Clericuzio, Antonio. 2010. "'Sooty Empiricks' and Natural Philosophy: The Status of Chemistry in the Seventeenth Century." *Science in Context*, 23: 329–50.

Clow, Archibald, and Nan L. Clow. 1952. *The Chemical Revolution: A Contribution to Social Technology*. London: Batchworth Press.

Coley, Noel G. 2001. "George Fordyce, M.D., F.R.S. (1736–1802): Physician-Chemist and Eccentric." *Notes and Records of the Royal Society of London*, 55: 395–409.

Contant, Jean-Paul. 1952. *L'enseignement de la chimie au Jardin Royal des Plantes de Paris*. Cahors: Coueslant.

Cook, Alexandra. 2016. "An Idea Ahead of Its Time: Jean-Jacques Rousseau's Mobile Botanical Laboratory." In Marianne Klemun and Ulrike Spring (eds), *Expeditions as Experiments: Practising Observation and Documentation*. Basingstoke: Palgrave Macmillan.

Cooper, Alix. 2007. *Inventing the Indigenous: Local Knowledge and Natural History in Early Modern Europe*. Cambridge: Cambridge University Press.

Cowen, David. L. 2001. *Pharmacopoeias and Related Literature in Britain and America, 1618–1847*. Farnham: Ashgate.

Craddock, Paul T. 2009. "The Origins and Inspirations of Zinc Smelting." *Journal of Materials Science*, 44: 2181–91.

Crawford, Matthew J. 2014. "An Empire's Extract: Chemical Manipulations of Cinchona Bark in the Eighteenth-Century Spanish Atlantic World." *Osiris*, 29: 215–29.

Crosland, Maurice. 1962. *Historical Studies in the Language of Chemistry*. London: Heinemann.

Crosland, Maurice. 1963. "The Development of Chemistry in the Eighteenth Century." *Studies of Voltaire and the Eighteenth Century*, 24: 369–441.

Crosland, Maurice. 1987. "The Image of Science as Threat: Burke versus Priestley and the 'Philosophic Revolution'." *British Journal for the History of Science*, 20: 277–307.

Crosland, Maurice. 1994. *In the Shadow of Lavoisier: The Annales de Chimie and the Establishment of a New Science*. Oxford: Alden Press.

Crosland, Maurice. 2000. "'Slippery Substances': Some Practical and Conceptual Problems in the Understanding of Gases in the Pre-Lavoisier Era." In Frederic L. Holmes and Trevor H. Levere (eds), *Instruments and Experimentation in the History of Chemistry*. Cambridge, MA: MIT Press.

Crosland, Maurice. 2009. "Lavoisier's Achievement: More Than a Chemical Revolution." *Ambix*, 56: 93–114.

Cullen, William. 1747. Lectures on Chemistry. Glasgow University Library, Cullen MSS, no. 7.

Cullen, William. n.d. Royal College of Physicians of Edinburgh, Cullen MSS, C11.

D'Alembert, Jean-Baptiste. [1751] 2009. "Discours préliminaire." In Denis Diderot and Jean-Baptiste D'Alembert (eds), *Encyclopédie ou Dictionnaire raisonné des sciences des arts et des métiers*. Vol. 1, i–xlv. English translation by Richard N. Schwab, Walter E. Rex. Ann Arbor, MI: Michigan Publishing, University of Michigan Library. Available online: http://hdl.handle.net/2027/spo.did2222.0001.083.

Dann, Georg Edmund. 1958. *Martin Heinrich Klaproth, 1743–1817*. Berlin: Akademie-Verlag.

Darling, A.S. 2002. "Non-ferrous metals." In Ian McNeil (ed), *An Encyclopedia of the History of Technology*. London: Routledge.

Darnton, Robert. 1987. *The Business of Enlightenment: A Publishing History of the Encyclopédie 1775–1800*. Cambridge, MA: Harvard University Press.

Dashkova, Ekaterina Romanovna. 1840. *Memoirs of the Princess Daschkaw: Lady of Honour to Catherine II, Empress of All the Russias*. Edited by W. Bradford. London: Colburn.

Davies, Gordon L. 1969. *The Earth in Decay: A History of British Geomorphology 1578–1878*. London: Macdonald.

Davy, Humphry. 1839. "A Discourse Introductory to a Course of Lectures on Chemistry, Delivered in the Theatre of the Royal Institution, on the 21st of January, 1802." In John Davy (ed.), *The Collected Works of Sir Humphry Davy*, vol. 2. London: Smith, Elder and Co.

Davy, René. 1955. *Contribution à l'étude des origine de la droguerie pharmaceutique et de l'industrie du sel ammoniac en France: l'apothicaire Antoine Baumé (1728–1804)*. Cahors: Couelsant.

Day, Joan. 1991. "Copper, Zinc and Brass Production." In Ronald F. Tylecote and Joan Day (eds), *The Industrial Revolution in Metals*. London: Institute of Metals.
Day, Lance. 2002. "The Chemical and Allied Industries." In Ian McNeil (ed.), *An Encyclopedia of the History of Technology*. London: Routledge.
Dean, Dennis. 1992. *James Hutton and the History of Geology*. Ithaca, NY: Cornell University Press.
Debus, Allen G. 1967. "Fire Analysis and the Elements in the Sixteenth and the Seventeenth Centuries." *Annals of Science*, 23: 127–47.
Debus, Allen G. 2001. *Chemistry and Medical Debate: Van Helmont to Boerhaave*. Canton, MA: Science History Publications.
De Luc, Jean-André. 1778–1780. *Lettres physiques et morales sur les montagnes et sur l'histoire de la terre et de l'homme*. Paris: La Haye.
Demeter, Tamás. 2012. "Hume's Experimental Method." *British Journal for the History of Philosophy*, 20: 577–99.
De Vos, Paula. 2007. "From Herbs to Alchemy: The Introduction of Chemical Medicine to Mexican Pharmacies in the Seventeenth and Eighteenth Centuries." *Journal of Spanish Cultural Studies*, 8: 135–68.
De Vries, Jan. 2008. *The Industrious Revolution: Consumer Behavior and the Household Economy, 1650 to the Present*. Cambridge: Cambridge University Press.
Desaguliers, John Theophilus. 1717. *Physico-Mechanical Lectures*. London: Bridger.
Diderot, Denis. [1753] 1964. *Pensées sur l'interprétation de la nature*. In Paul Vernière (ed.), *Oeuvres philosophiques*. Paris: Garnier.
Dolan, Brian. 1998. "Transferring Skill: Blowpipe Analysis in Sweden and England, 1750–1850." In Brian Dolan (ed.), *Science Unbound: Geography Space and Discipline*. Umeå: Umeå Univeriteit Skrifter.
Dolan, Brian. 2004. *Wedgwood: The First Tycoon*. New York, NY: Viking.
Dolin, Eric Jay. 2007. *Leviathan: The History of Whaling in America*. New York, NY: Norton.
Donovan, Arthur. 1975. *Philosophical Chemistry in the Scottish Enlightenment*. Edinburgh: Edinburgh University Press.
Donovan, Arthur (ed.). 1988. *The Chemical Revolution: Essays in Reinterpretation*. Philadelphia, PA: University of Pennsylvania Press.
Donovan, Arthur, and Joseph Prentiss. 1980. "James Hutton's Medical Dissertation." *Transactions of the American Philosophical Society*, 70: 3–57.
Dossie, Robert. 1758. *The Elaboratory Laid Open, or, The Secrets of Modern Chemistry and Phramacy Revealed*. London: Nourse.
Duhem, Pierre. [1902] 2002. *Mixture and Chemical Combination, and Related Essays*. Edited and translated with an introduction by Paul Needham. Dordrecht: Kluwer.
Duncan, Alistair. 1996. *Laws and Order in Eighteenth-Century Chemistry*. Oxford: Clarendon Press.
Earl, Bryan. 1991. "Tin preparation and smelting." In Ronald F. Tylecote and Joan Day (eds), *The Industrial Revolution in Metals*. London: Institute of Metals.
Eddy, Matthew Daniel. 2007. "The Aberdeen Agricola: Chemical Principles and Practice in James Anderson's Georgics and Geology." In Lawrence M. Principe (ed.), *New Narratives in Eighteenth-Century Chemistry*. Dordrecht: Springer.
Eddy, Matthew Daniel. 2008. *The Language of Mineralogy: John Walker, Chemistry and the Edinburgh Medical School, 1750–1800*. London: Routledge.
Eddy, Matthew Daniel. 2010. "The Sparkling Nectar of Spas; or, Mineral Water as a Medically Commodifiable Material in the Province, 1770–1805." In Ursula Klein

and Emma C. Spary (eds), *Materials and Expertise in Early Modern Europe: Between Market and Laboratory*. Chicago, IL: University of Chicago Press.

Eddy, Matthew Daniel. 2014. "How to See a Diagram: A Visual Anthropology of Chemical Affinity." *Osiris*, 26: 178–96.

Eddy, Matthew Daniel. 2022. *Media and the Mind: Art, Science and Notebooks as Paper Machines, 1700–1830*. Chicago, IL: University of Chicago Press.

Eddy, Matthew Daniel, Seymour H. Mauskopf, and William R. Newman. 2014. "An Introduction to Chemical Knowledge in the Early Modern World." In Matthew Daniel Eddy, Seymour H. Mauskopf, and William R. Newman (eds), *Chemical Knowledge in the Early Modern World (Osiris)*, 29: 1–15. Chicago, IL: University of Chicago Press Journals.

Egerton, Frank N. 2012. *Roots of Ecology: Antiquity to Haeckel*. Berkeley, CA: University of California Press.

Eklund, Jon. 1975. *The Incompleat Chymist: Being an Essay on the Eighteenth-Century Chemist in his Laboratory. with a Dictionary of Obsolete Chemical Terms of the Period*. Washington, DC: Smithsonian Institution Press.

Emerton, Norma E. 1984. *The Scientific Reinterpretation of Form*. Ithaca, NY: Cornell University Press.

État de medicine, chirurgie et pharmacie en France, en Europe pour l'année 1776. 1776. Paris: Didot.

Fairlie, Susan. 1965. "Dyestuffs in the Eighteenth Century." *Economic History Review*, 17: 488–510.

Fester, Gustav A. 1923 [1969]. *Die Entwicklung der chemischen Technik bis zu den Anfängen der Grossindustrie: Ein technologisch-historischer Versuch*. Wiesbaden: Sändig.

Fontenelle, Bernard. [1686] 2005. *Entretiens sur la pluralité des mondes habités*. La Tour d'Aigues: L'aube.

Fontenelle, Bernard. 1719. "Eloge de Nicolas Lémery." *Histoire de l'Académie royale des sciences, année 1715*. Amsterdam. 96–108.

Forbes, Robert J. 1970. *A Short History of the Art of Distillation*. Leiden: Brill.

Fors, Hjalmar. 2003. *Mutual Favours: The Social and Scientific Practice of Eighteenth-Century Swedish Chemistry*. Uppsala: Universiteit Stryckereeit.

Fors, Hjalmar. 2015. *The Limits of Matter: Chemistry, Mining and Enlightenment*. Chicago, IL: University of Chicago Press.

Fourcroy, Antoine-François. 1805. "Laboratoire." In *Encyclopédie Méthodique – Chymie*, vol. 4. Paris: Agasse.

Frercks, Jan. 2010. "Demonstrating the Facticity of Facts: University Lectures in Chemistry as Science in Germany around 1800." *Ambix*, 57: 65–83.

Frercks, Jan, and Michael Markert. 2007. "The Invention of *Theoretische Chemie*: Forms and Uses of German Chemistry Textbooks, 1775–1820." *Ambix*, 54: 130–55.

Fries, Theodor M. 1923. *Linnaeus*. London: Witherby.

Fussell, George E. 1969. "Science and Practice in Eighteenth-Century British Agriculture." *Agricultural History*, 43: 7–18.

Gaukroger, Stephen. 2010. *The Collapse of Mechanism and the Rise of Sensibility: Science and the Shaping of Modernity, 1680–1760*. Oxford: Oxford University Press.

Geoffroy, Etienne-François. 1718. "Table des differents rapports observés en chimie entre differentes substances." *Memoires de l'Academie Royale des Sciences*, 202–12. Paris: Imprimerie Royale.

Geoffroy, Etienne-François. [1718] 1996. "Table of the Different Relations Observed in Chemistry Between Different Substances." *Science in Context*, 9: 313–20.

Gibbs, Frederick W. 1939. "The History of the Manufacture of Soap." *Annals of Science*, 4: 169–90.

Gibbs, Frederick W. 1951. "Peter Shaw and the Revival of Chemistry." *Annals of Science*, 7: 212–32.

Gibbs, Frederick W. 1952. "William Lewis, M.B., F.R.S. (1708–1781)." *Annals of Science*, 8: 122–51.

Gibbs, Frederick W. 1953. "George Wilson (1631–1711)." *Endeavour*, 12: 182–5.

Gibbs, Frederick W. 1958. "Boerhaave's Chemical Writings." *Ambix*, 6: 117–35.

Gillispie, Charles. 2004. *Science and Polity in France: The Revolutionary and Napoleonic Years*. Princeton, NJ: Princeton University Press.

Gillray, James. 1790. *Smelling Out a Rat; – or – The Atheistical-Revolutionist Disturbed in His Midnight "Calculations."* London: Humphrey.

Gillray, James. 1796. *The Dissolution; or – The Alchymist Producing an Aetherial Representation*. London: Humphrey.

Gittins, Lawrence. 1979. "Innovations in Textile Bleaching in Britain in the Eighteenth Century." *Business History Review*, 53: 194–204.

Glacken, Clarence J. 1976. *Traces on the Rhodian Shore: Nature and Culture in Western Thought from Ancient Times to the End of the Eighteenth Century*. Berkeley, CA: University of California Press.

Gleeson, Janet. 2013. *The Arcanum*. London: Random House.

Goethe, Johann Wolfgang von. [1809a] 1900. *Elective Affinities*. New York, NY: Collier.

Goethe, Johann Wolfgang von. 1809b. "Advertisement." *Morgenblatt für gebildete Stände*, 4 September.

Goethe, Johann Wolfgang von. [1809] 1999. *Elective Affinities*. Oxford: Oxford University Press.

Golinski, Jan. 1983. "Peter Shaw: Chemistry and Communication in Augustan England." *Ambix*, 30: 19–29.

Golinski, Jan. 1992. *Science as Public Culture: Chemistry and Enlightenment in Britain, 1760–1820*. Cambridge: Cambridge University Press.

Golinski, Jan. 1995. "'The Nicety of Experiment': Precision of Measurement and Precision of Reasoning in Late Eighteenth-Century Chemistry." In Norton Wise (ed.), *The Values of Precision*. Princeton, NJ: Princeton University Press.

Golinski, Jan. 1999. *Science as Public Culture. Chemistry and Enlightenment in Britain, 1760–1820*. Cambridge: Cambridge University Press.

Golinski, Jan. 2003. "Chemistry." In Roy Porter (ed.), *The Cambridge History of Science*, vol. 4. Cambridge: Cambridge University Press.

Golinski, Jan. 2010. *British Weather and the Climate of Enlightenment*. Chicago, IL: University of Chicago Press.

Golinski, Jan. 2017. "Sublime Astronomy: The Eidouranion of Adam Walker and His Sons." *Huntington Library Quarterly*, 80: 135–57.

Grapi, Pere. 2001. "The Marginalization of Berthollet's Chemical Affinities in the French Textbook Tradition at the Beginning of the Nineteenth Century." *Annals of Science*, 58: 115–36.

Grison, Emmanuel, Patrice Bert, and Michelle Goupil (eds). 1995. *La Correspondance entre Kirwan et Guyton de Morveau (1782–1802)*. Berkeley, CA: University of California Press.

Guerlac, Henry. 1956. "A Note on Lavoisier's Scientific Education." *Isis*, 47: 211–16.
Guerlac, Henry. 1957a. "Joseph Black and Fixed Air." *Isis*, 48: 124–51.
Guerlac, Henry. 1957b. "Joseph Black and Fixed Air: Part II." *Isis*, 48: 433–56.
Guerlac, Henry. 1959. "Some French Antecedents of the Chemical Revolution." *Chymia*, 5: 73–112.
Guerlac, Henry. 1961. *Lavoisier – The Crucial Year. The Background and Origin of his First Experiments on Combustion*. Ithaca, NY: Cornell University Press.
Guerlac, Henry. 1976. "Chemistry as a Branch of Physics: Laplace's Collaboration with Lavoisier." *Historical Studies in the Physical Sciences*, 7: 193–276.
Guerlac, Henry. 1981. "Lavoisier, Antoine-Laurent." In Charles Coulston Gillispie (ed.), *Dictionary of Scientific Biography*, vol. 8. New York, NY: Charles Scribner's Sons.
Guerlac, Henry. 1982. "Joseph Black's Work on Heat." In A.D.C. Simpson (ed.), *Joseph Black, 1728–1799: A Commemorative Symposium*. Edinburgh: Royal Scottish Museum.
Guerra, Corinna. 2015. "If You Don't Have a Laboratory Find a Good Volcano: Mount Vesuvius as a Natural Chemical Laboratory in Eighteenth-Century Italy." *Ambix*, 62: 245–65.
Guerrini, Anita. 1994. "Chemistry Teaching at Oxford and Cambridge, circa 1700." In P. Rattansi and A. Clericuzio (eds), *Alchemy and Chemistry in the Sixteenth and Seventeenth Centuries*. Dordrecht: Kluwer.
Guillerme, André. 2007. *La naissance de l'industrie à Paris: Entre sueurs et vapeurs: 1780–1830*. Seysse: Champ Vallon.
Hales, Stephen. 1727. *Vegetable Staticks: Or, An Account of some Statical Experiments on the Sap in Vegetables: Being an Essay towards a Natural History of Vegetation*. London: Innys.
Hales, Stephen. 1738. *Statical Essays: Containing Vegetable Staticks*. The Third Edition, with Amendments. London: Innys and Manby.
Hales, Stephen. 1743. *A Description of Ventilators*. London: Innys.
Hall, A. Rupert. 1967. "Scientific Method and the Progress of Techniques." In *The Cambridge Economic History of Europe: Vol. 4, The Economy of Expanding Europe in the Sixteenth and Seventeenth Centuries*. Cambridge: Cambridge University Press.
Hannaway, Owen. 1975. *The Chemists and the Word: The Didactic Origins of Chemistry*. Baltimore, MD: Johns Hopkins University Press.
Heilbron, John. 1993. "Weighing Imponderables and Other Quantitative Science around 1800." *Historical Studies in the Physical and Biological Sciences*, 24: 1–33, 35–277, 279–337.
Hendry, Robin Findlay. 2019. "Elements and (First) Principles in Chemistry." *Synthese*, open access online: https://doi.org./10.1007/s11229-019-02312-8.
Hermbstädt, Sigismund Friedrich. 1808. *Grundsätze der experimentellen Kammeral-Chemie*. Berlin: Realschulbuchhandlung.
Heyl, Christoph. 2013. "William Hogarth, Science and Human Nature." In Ralf Haekel and Sabine Blackmore (eds), *Discovering the Human*. Göttingen: Vandenhoek & Ruprecht.
Hjelm, Peter Jacob. 1786. *Åminnelse-tal öfver... Torbern Olof Bergman*. Stockholm: Lange.
Hobbs, William. 1981. *The Earth Generated and Anatomized: An Early Eighteenth Century Theory of the Earth*. London: British Museum.

Hoffmann, Gottfried A. 1779. *Anleitung zur Chemie für Künstler und Fabrikanten*. Gotha: Ettinger.
Holden, Norman E. 2007. "Chemical Elements: History of the Origin." In Glenn Considine and Peter Kulik (eds), *Van Nostrand's Scientific Encyclopedia*. New York, NY: Wiley.
Holmes, Frederic Lawrence. 1962. "From Elective Affinity to Chemical Equilibrium: Berthollet's Laws of Mass Action." *Chymia*, 8: 105–45.
Holmes, Frederic Lawrence. 1971. "Analysis by Fire and Solvent Extractions: The Metamorphosis of a Tradition." *Isis*, 62: 128–48.
Holmes, Frederic Lawrence. 1984. *Lavoisier and the Chemistry of Life: An Exploration of Scientific Creativity*. Madison, WI: University of Wisconsin Press.
Holmes, Frederic Lawrence. 1988. "Lavoisier's Conceptual Passage." *Osiris*, 4: 82–92.
Holmes, Frederic Lawrence. 1989. *Eighteenth-Century Chemistry as an Investigative Enterprise*. Berkeley, CA: Office for the History of Science and Technology, University of California.
Holmes, Frederic Lawrence. 1996. "The Communal Context for Etienne-François Geoffroy's 'Table des rapports'." *Science in Context*, 9: 289–311.
Holmes, Frederic Lawrence. 1997. *Antoine Lavoisier: The Next Crucial Year, Or, The Sources of His Quantitative Method in Chemistry*. Princeton, NJ: Princeton University Press.
Holmes, Frederic Lawrence. 2004. "Investigation and Pedagogical Style in French Chemistry at the End of the 17th Century." *History of Science*, 34: 277–309.
Holmes, Frederic Lawrence, and Trevor H. Levere (eds). 2000. *Instruments and Experimentation in the History of Chemistry*. Cambridge, MA: MIT Press.
Homburg, Ernst, and Johan H. De Vlieger. 1996. "A Victory of Practice over Science: The Unsuccessful Modernisation of the Dutch White Lead Industry (1780–1865)." *History and Technology*, 13: 33–52.
Howard, Rio. 1981. "Guy de la Brosse and the *Jardin des Plantes* in Paris." In Harry Woolf (ed.), *The Analytic Spirit: Essays in the History of Science in Honor of Henry Guerlac*. Ithaca, NY: Cornell University Press.
Hufbauer, Karl. 1982. *The Formation of the German Chemical Community (1720–1795)*. Berkeley, CA: University of California Press.
Hyde, Charles K. 1977. *Technological Change and the British Iron Industry, 1700–1870*. Princeton, NJ: Princeton University Press.
Irwin, Emily. 2012. "The Spermaceti Candle and the American Whaling Industry." *Historia*, 21: 45–53.
Jacob, Margaret C., and Larry Stewart. 2004. *Practical Matter: Newton's Science in the Service of Industry and Empire, 1687–1851*. Cambridge, MA: Harvard University Press.
Jankovic, Vladimir. 2010. *Confronting the Climate: British Airs and the Making of Environmental Medicine*. Berlin: Springer.
Jay, Mike. 2009. *The Atmosphere of Heaven: The Unnatural Experiments of Dr. Beddoes and His Sons of Genius*. New Haven, CT: Yale University Press.
Jensen, William B. 1986. "The Development of Blowpipe Analysis." In John Stock and Mary V. Orna (eds), *The History and Preservation of Chemical Instrumentation*. Dordrecht: Kluwer.
Johns, Adrian. 1998. *The Nature of the Book: Print and Knowledge in the Making*. Chicago, IL: University of Chicago Press.
Joly, Bernard. 2014. "Etienne-François Geoffroy (1672–1731), a Chemist on the Frontiers." *Osiris*, 29: 117–31.

Jones, Peter M. 2016. "Making Chemistry the 'Science' of Agriculture, c. 1760–1840." *History of Science*, 54: 169–94.
Jungnickel, Christa, and Russell McCormmach. 1999. *Cavendish: The Experimental Life*. Lewisburg, PA: Bucknell.
Kim, Mi Gyung. 2003. *Affinity, That Elusive Dream: A Genealogy of the Chemical Revolution*. Cambridge, MA: MIT Press.
Kim, Mi Gyung. 2017. *The Imagined Empire: Balloon Enlightenments in Revolutionary Europe*. Pittsburgh, PA: University of Pittsburgh Press.
Klein, Ursula. 1994a. *Verbindung und Affinität. Die Grundlegung der neuzeitlichen Chemie an der Wende des 17. zum 18. Jahrhundert*. Basel: Birkhäuser.
Klein, Ursula. 1994b. "Origin of the Concept of Chemical Compound." *Science in Context*, 7: 163–204.
Klein, Ursula. 1995. "E.F. Geoffroy's Table of Different 'Rapports' Observed between Different Chemical Substances – A Reinterpretation." *Ambix*, 42: 79–100.
Klein, Ursula. 1996. "The Chemical Workshop Models and the Experimental Practice: Continuities and Discontinuities." *Science in Context*, 9: 251–87.
Klein Ursula. 2005a. "Shifting Ontologies, Changing Classifications: Plant Materials from 1700 to 1830." *Studies in History and Philosophy of Science*, 36: 261–329.
Klein, Ursula. 2005b. "Contexts and Limits of Lavoisier's Analytical Plant Chemistry: Plant Materials and their Classification." *Ambix*, 52: 107–57.
Klein, Ursula. 2005c. "Technoscience Avant la Lettre." *Perspectives on Science*, 13: 1–48.
Klein, Ursula. 2007. "Apothecary-Chemists in Eighteenth-Century Germany." In Lawrence M. Principe (ed.), *New Narratives in Eighteenth Century Chemistry*. Dordrecht: Springer.
Klein, Ursula 2008. "The Laboratory Challenge: Some Revisions of the Standard Picture of Early Modern Experimentation." *Isis*, 99: 769–82.
Klein, Ursula. 2009. "Chemical and Pharmaceutical Laboratories before the Professionalization of Chemistry." In Marta Lourenço and Ana Carneiro (eds), *Spaces and Collections in the History of Science*. Lisbon: Museum of Science of the University of Lisbon.
Klein Ursula. 2010. "Blending Technical Innovation and Learned Natural Knowledge: The Making of Ethers." In Ursula Klein and Emma C. Spary (eds), *Materials and Expertise in Early Modern Europe: Between Market and Laboratory*. Chicago, IL: University of Chicago Press.
Klein, Ursula. 2012a. "The Prussian Mining Official Alexander von Humboldt." *Annals of Science*, 69: 27–68.
Klein, Ursula. 2012b. "Artisanal-Scientific Experts in Eighteenth-Century France and Germany." Special issue of *Annals of Science*, 69: 303–433.
Klein, Ursula. 2012c. "Savant Officials in the Prussian Mining Administration." *Annals of Science*, 69: 349–74.
Klein, Ursula. 2013. "Chemical Experts at the Royal Prussian Porcelain Manufactory." *Ambix*, 60: 99–121.
Klein, Ursula. 2014a. "Chemical Expertise: Chemistry in the Royal Prussian Porcelain Manufactory." *Osiris*, 29: 262–82.
Klein, Ursula. 2014b. "Klaproth's Discovery of Uranium." In Ursula Klein and Carstein Reinhardt (eds), *Objects of Chemical Inquiry*. Sagamore Beach, MA: Science History Publications.
Klein, Ursula. 2014c. "Depersonalizing the Arcanum." *Technology and Culture*, 55: 591–621.

Klein, Ursula. 2015a. *Humboldts Preußen: Wissenschaft und Technik im Aufbruch*. Darmstadt: Wissenschaftliche Buchgesellschaft.
Klein, Ursula. 2015b. "A Revolution That Never Happened." *Studies in History and Philosophy of Science*, 49: 80–90.
Klein, Ursula. 2017. "Hybrid Experts." In Matteo Valleriani (ed.), *The Structures of Practical Knowledge*. Cham: Springer.
Klein, Ursula. 2020. *Technoscience in History, Prussia, 1750–1850*. Cambridge, MA: MIT Press.
Klein, Ursula, and Wolfgang Lefèvre. 2007. *Materials in Eighteenth-Century Science, A Historical Ontology*. Cambridge, MA: MIT Press.
Klein, Ursula, and Emma C. Spary (eds). 2010. *Materials and Expertise in Early Modern Europe: Between Market and Laboratory*. Chicago, IL: University of Chicago Press.
Knight, David M. 1986. "Accomplishment or Dogma: Chemistry in the Introductory Works of Jane Marcet and Samuel Parkes." *Ambix*, 33: 94–98.
Knight, David M. 1992. *Ideas in Chemistry: A History of the Science*. London: Athlone.
Knight, David M. 2004. *Science and Spirituality: The Volatile Connection*. London: Routledge.
Knight, David M. 2006. "Popularizing Chemistry: Hands-On and Hands-Off." *Hyle: International Journal for Philosophy of Chemistry*, 12: 131–40.
Knight, David M. 2013. "Chemical Sciences and Natural Theology." In Russell Re Manning (ed.), *The Oxford Handbook of Natural Theology*. Oxford: Oxford University Press.
Knight, David M. 2016. *Science in the Romantic Era*. London: Routledge.
Knoeff, Rina. 2007. "Practicing Chemistry 'After the Hippocratical Manner'." In Lawrence M. Principe (ed.), *New Narratives in Eighteenth-Century Chemistry*. Dordrecht: Springer.
Köhler, Ulrike Kristina. 2013. "Ann Radcliff's Gothic – A Subtile Plea for Female Education in the Arts and the Sciences." In Ralf Haekel and Sabine Blackmore (eds), *Discovering the Human: Life Science and the Arts in the Eighteenth and Early Nineteenth Centuries*. Göttingen: V&R Unipress.
Konecný, Peter. 2012. "The Hybrid Expert in the 'Bergstaat': Anton von Ruprecht as a Professor of Chemistry and Mining and as a Mining Official, 1779–1814." *Annals of Science*, 69: 335–47.
Krätz, Otto. 1999. *Goethe und die Naturwissenschaften*. Munich: Callwey.
Kraft, Alexander. 2008. "On the Discovery and History of Prussian Blue." *Bulletin for the History of Chemistry*, 33 (2): 61–67.
Kremers, Edward, and George Urdang. 1976. *History of Pharmacy: A Guide and Survey*. Philadelphia, PA: Lippincott.
Krohn, Wolfgang, and Wolf Schäfer. 1983. "Agricultural Chemistry: The Origin and Structure of a Finalized Science." In Wolf Schäfer (ed.), *Finalization in Science*. Dordrecht: Springer.
Kumar, Prakash. 2012. *Indigo Plantations and Science in Colonial India*. Cambridge: Cambridge University Press.
Lanoé, Catherine. 2003. "Les jeux de l'artificiel: Culture, production et consommation des cosmétiques à Paris sous l'Ancien Régime, XVIe–XVIIIe siècles." PhD thesis, Université Paris-I Panthéon-Sorbonne.
Laudan, Rachel. 1987. *From Mineralogy to Geology: The Foundations of a Science, 1650–1830*. Chicago, IL: University of Chicago Press.

Lavoisier, Antoine Laurent. 1774. *Opuscules physiques et chymiques*. Paris: Durand.
Lavoisier, Antoine Laurent. 1789. *Traité élémentaire de chimie*. Paris: Cuchet.
Lavoisier, Antoine Laurent. [1789] 1965. *Elements of Chemistry, translated by Robert Kerr, Edinburgh, 1790*. New York, NY: Dover.
Lavoisier, Antoine Laurent. 1790. *Elements of Chemistry, in a New Systematic Order, Containing All the Modern Discoveries*, trans. R. Kerr. Edinburgh: W. Creech.
Lavoisier, Marie-Anne-Pierrette. 1790. *Antoine Lavoisier Performing a Respiration Experiment* (ca. 1790). London: Wellcome Library.
Lefèvre, Wolfgang. 2018. "The Méthode de nomenclature chimique (1787): A Document of Transition." *Ambix*, 65: 9–29.
Lehman, Christine. 2008. "Between Commerce and Philanthropy: Chemistry Courses in Eighteenth Century Paris." In Bernadette Bensaude-Vincent and Christine Blondel (eds), *Science and Spectacle in the European Enlightenment*. Aldershot: Ashgate.
Lehman, Christine. 2009a. "Les deux faces de la chimie de Venel: côté cours, côté Encyclopédie." *Corpus*, 56: 87–116.
Lehman, Christine. 2009b. "Mid-Eighteenth Century Chemistry in France as Seen Through Student Notes from the Courses of Gabriel-François Venel and Guillaume-François Rouelle." *Ambix*, 56: 163–89.
Lehman, Christine. 2010. "Innovation in Chemistry Courses in France in the Mid-Eighteenth Century: Experiments and Affinities. *Ambix*, 57: 3–26.
Lehman, Christine. 2012. "Pierre-Joseph Macquer: An Eighteenth-Century Artisanal Scientific Expert." *Annals of Science*, 69: 307–33.
Lehman, Christine. 2013. "Alchemy Revisited by the Mid-Eighteenth Century Chemists in France: An Unpublished Manuscript by Pierre-Joseph Macquer." *Nuncius*, 28: 156–216.
Lehman, Christine. 2014. "Pierre-Joseph Macquer: Chemistry in the French Enlightenment." *Osiris*, 29: 245–61.
Lehman, Christine. 2016. "What is the True Nature of Diamond?" *Nuncius*, 31: 361–407.
Lehman, Christine, and François Pepin. 2009. "La chimie et l'*Encyclopédie*." *Corpus: Revue de philosophie*, 56.
Leigh, G. Jeffery, and Alan J. Rocke. 2016. "Women and Chemistry in Regency England: New Light on the Marcet Circle." *Ambix*, 63: 28–45.
Lémery, Nicolas. 1677. *A Course of Chymistry: Containing The Easiest Manner of performing those Operations that are in Use in Physick*. London: Kettilby.
Lémery, Nicolas. 1683. *Cours de chymie, contenant la manière de faire les opérations qui sont en usage dans la médecine, par une méthode facile*. Paris.
Lémery, Nicolas. 1713. *Cours de chymie, contenant la manière de faire les opérations qui sont en usage dans la médecine, par une méthode facile*. Paris.
Lémery, Nicolas. 1720. *A Course of Chymistry: Containing An Easie Method of Preparing those Chymical Medicines Which are used in Physick, the fourth edition*. London: Bell.
Lémery, Nicolas. 1724. *Cours de chymie, contenant la manière de faire les opérations qui sont en usage dans la médecine par une méthode facile. Douzième édition*. Lyon: Guerrier.
Lémery, Nicolas. 1757. *Cours de chymie contenant la manière de faire les opérations qui sont en usage dans la médecine, par une méthode facile. Nouvelle édition, revue, corrigée & augmentée*. Paris: d'Houry.
Leong, Elaine. 2008. "Making Medicines in the Early Modern Household." *Bulletin of the History of Medicine*, 82: 145–68.

Leong, Elaine. 2018. *Recipes and Everyday Knowledge: Medicine, Science, and the Household in Early Modern England*. Chicago, IL: University of Chicago Press.

Lepenies, Wolf. 1982. "Linnaeus's *Nemesis Divina* and the Concept of Divine Retaliation." *Isis*, 73: 11–27.

Leqan, Mai. 2000. *La chimie selon Kant*. Paris: Presses Universitaires de France.

Leqan, Mai. 2010. "D'une 'philosophie de la chimie' au chimisme en philosophie: Les premiers écrits de Schelling sur la nature." *Dix-huitième siècle*, 42: 491–512.

Le Roux, Thomas. 2017. "Between Industry and the Environment: Chemical Governance in France, 1770–1830." In Lissa Roberts and Simon Werrett (eds), *Compound Histories: Materials, Governance and Production, 1760–1840*. Leiden: Brill.

Le Sage, René. [1707] 1841. *Le diable boiteux*. Translated by Joseph Thomas as *The Devil on Two Sticks*. London: Hutchinson.

Levere, Trevor H. 1990. "Lavoisier: Language, Instruments and the Chemical Revolution." In Trevor H. Levere and William R. Shea (eds), *Experiment, and the Sciences: Essays on Galileo and the History of Science in Honour of Stillman Drake*. Dordrecht: Kluwer.

Levere, Trevor H. 1994. *Chemists and Chemistry in Nature and Society 1770–1878*. Aldershot: Ashgate.

Levere, Trevor H. 2000. "Measuring Gases and Measuring Goodness." In Frederic Lawrence Holmes and Trevor H. Levere (eds), *Instruments and Experimentation in the History of Chemistry*. Cambridge, MA: MIT Press.

Levere, Trevor H. 2001. *Transforming Matter: A History of Chemistry from Alchemy to the Buckyball*. Baltimore, MD: Johns Hopkins University Press.

Levere, Trevor H. 2005. "Lavoisier's Gasometers and Others: Research, Control, and Dissemination." In Marco Beretta (ed.), *Lavoisier in Perspective*. Munich: Deutsches Museum.

Levere, Trevor H., and Gerald L'E. Turner (eds). 2002. *Discussing Chemistry and Steam: The Minutes of a Coffee House Philosophical Society 1780–1787*. Oxford: Oxford University Press.

Lewis, William. 1763. *Commercium Philosophico-Technicum*. London: Baldwin.

Licoppe, Christian. 1996. *La formation de la pratique scientifique: Le discours de l'expérience en France et en Angleterre (1630–1820)*. Paris: Editions la Découverte.

Lindeboom, G. A. 1973. "David en Nicholaas Stam, apothekers te Leiden," *Pharmaceutisch Weekblad*, 108: 153–60.

Love, Rosaleen. 1974. "Herman Boerhaave and the Instrument-Element Concept of Fire." *Ambix*, 31: 547–59.

Lowengard, Sarah. 2008. *The Creation of Color in Eighteenth-Century Europe*. New York, NY: Columbia University Press.

Lynn, Michael R. 2006. *Popular Science and Public Opinion in Eighteenth-Century France*. Manchester: Manchester University Press.

Lyon, John. 1781. *Farther Proofs that Glass is Permeable by the Electric Effluvia*. London: Dodsley.

Macquer, Pierre-Joseph. 1749. *Elémens de chymie théorique*. Paris: Hérissant.

Macquer, Pierre-Joseph. 1753. *Elémens de chymie-théorique*. Paris: Hérrisant.

Macquer, Pierre-Joseph. 1758. *Elements of the Theory and Practice of Chemistry*. London: Millar and Nourse.

Macquer, Pierre-Joseph. 1766. *Dictionnaire de chymie*. 2 vols. Paris: Lacombe.

Macquer, Pierre-Joseph. [1766] 1777. *A Dictionary of Chemistry*. English translation by James Keir. London: Cadell and Elmsly.

Macquer, Pierre-Joseph. 1778. *Dictionnaire de chymie*. 2 vols., 2nd ed. Paris: Barrois.

Malaquias, Isabel. 2008. "Aspects of John Hyacinth de Magellan's Scientific Network." In José Bertomeu-Sanchez, Duncan Thorburn Burns, and Brigitte Van Tiggelen (eds), *Neighbors and Territories: The Evolving Identity of Chemistry*. Louvain-la-Neuve: Mémosciences.

Martinón-Torres, Marcos, Thilo Rheren, and Ian C. Freestone. 2006. "Mullite and the Mystery of the Hessian Wares." *Nature*, 444: 437–38.

Mauskopf, Seymour H. 1976. *Crystals and Compounds: Molecular Structure and Composition in Nineteenth-Century French Science*. Philadelphia, PA: American Philosophical Society.

Mauskopf, Seymour H. 1995. "Lavoisier and the Improvement of Gunpowder Production." *Revue d'histoire des sciences*, 48: 95–121.

Mauskopf, Seymour. 2007. "Reflections: A Likely Story." In Lawrence Principe (ed.), *New Narratives in Eighteenth-Century Chemistry*. Dordrecht: Springer.

Mauskopf, Seymour H. 2010. "The Crisis of English Gunpowder in the Eighteenth Century." In Ursula Klein and Emma Spary (eds), *Materials and Expertise in Early Modern Europe: Between Market and Laboratory*. Chicago, IL: University of Chicago Press.

McArthur, James. 1801. *Financial and Political Facts of the Eighteenth Century*, 3rd ed. London: Wright.

McEvoy, John G. 2010. *The Historiography of the Chemical Revolution: Patterns of Interpretation on the History of Science*. London: Pickering and Chatto.

Meinel, Christoph. 1983. "Theory or Practice? The Eighteenth-Century Debate on the Scientific Status of Chemistry." *Ambix*, 30: 121–32.

Meinel, Christoph. 1988. "'Artibus Academicis Inserenda': Chemistry's Place in Eighteenth and Early Nineteenth-Century Universities." *History of Universities*, 8: 89–115.

Mercier, Louis-Sébastien. [1781] 1999. *Panorama of Paris: Selections from Tableau de Paris*. English trans. of selections from *Tableau de Paris* (1781). University Park: Pennsylvania State University Press.

Metzger, Hélène. [1930] 2006. *Hélène Metzger's Newton, Stahl, Boerhaave, and Chemical Doctrine*. Hamilton, Ontario: Huxley.

Miller, David Philip. 2004. *Discovering Water: James Watt, Henry Cavendish and the Nineteenth-century Water Controversy*. Aldershot: Ashgate.

Moilliet, J.L. 1964. "Keir's 'Dialogues on Chemistry' – An Unpublished Masterpiece." *Chemistry and Industry*, 101: 2081–83.

Moran, Bruce T. 1996. "A Survey of Chemical Medicine in the 17th Century: Spanning Court, Classroom, and Cultures." *Pharmacy in History*, 38: 121–33.

Morrell, Jack B. 1971. "Professors Robison and Playfair, and the 'Theophobia Gallica': Natural Philosophy, Religion and Politics in Edinburgh, 1789–1815." *Notes and Records of the Royal Society*, 26: 43–63.

Morris, Peter J.T. 2015. *The Matter Factory: A History of the Chemistry Laboratory*. London: Reaktion.

Moulden, John C. 1916. "Zinc, its Production and Industrial Applications." *Journal of the Royal Society of Arts*, 64: 495–513.

Multhauf, Robert P. 1966. *The Origins of Chemistry*. London: Oldbourne.

Musson, Albert Edward, and Eric Robinson. 1969. *Science and Technology in the Industrial Revolution*. Manchester: Manchester University Press.

Neville, Roy G., and William A. Smeaton. 1981. "Macquer's *Dictionnaire de Chymie*: A Bibliographical Study." *Annals of Science*, 38: 613–62.

Newman, William R. 2006. *Atoms and Alchemy: Chymistry and the Experimental Origins of the Scientific Revolution*. Chicago, IL: University of Chicago Press.

Newman, William R. 2014. "Robert Boyle, Transmutation, and the History of Chemistry before Lavoisier: A Response to Kuhn." *Osiris*, 29: 63–77.

Newman, William R., and Lawrence Principe. 1998. "Alchemy *vs* Chemistry: The Etymological Origins of a Historiographic Mistake." *Early Modern Science and Medicine*, 3: 32–65.

Newman, William R., and Lawrence M. Principe. 2002. *Alchemy Tried in the Fire: Starkey, Boyle, and the Fate Of Helmontian Chymistry*. Chicago, IL: University of Chicago Press.

Nieto-Galan, Agustí. 2001. *Colouring Textiles. A History of Natural Dyestuffs in Industrial Europe*. Dordrecht: Kluwer.

Niinistö, L. 1990. "Analytical Instrumentation in the 18th Century." *Fresenius' Journal of Analytical Chemistry*, 337: 213–17.

Oldroyd, David. 1973. "An Examination of G. E. Stahl's *Philosophical Principles of Universal Chemistry*." *Annals of Science*, 31: 36–52.

Orland, Barbara. 2010. "Enlightened Milk: Reshaping a Bodily Substance into a Chemical Object." In Ursula Klein and Emma C. Spary (eds), *Materials and Expertise in Early Modern Europe: Between Market and Laboratory*. Chicago, IL: University of Chicago Press.

Parascandola, John, and Aaron J. Ihde. 1969. "History of the Pneumatic Trough." *Isis*, 60: 351–61.

Partington, James R. 1961. *A History of Chemistry*, vol. 2. London: Macmillan.

Partington, James R. 1962. *A History of Chemistry*, vol. 3. London: Macmillan.

Patterson, Thomas S. 1937. "Jean Beguin and his *Tyrocinium Chymicum*." *Annals of Science*, 2: 243–98.

Pépin, François. 2013. "La nature naturante et les puissances de la matière, ou comment penser l'immanence totale à partir de la chimie." *Dix-huitième siècle*, 45: 131–48.

Percival, Thomas. 1769. *Experiments and Observations on Water Particularly on the Hard Pump Water of Manchester*. London: Johnson.

Perkins, John. 2003. "Creating Chemistry in Provincial France before the Revolution: The Example of Nancy and Metz. Part 1. Nancy." *Ambix*, 50: 145–81.

Perkins, John. 2004. "Creating Chemistry in Provincial France before the Revolution: The Example of Nancy and Metz. Part 2. Metz." *Ambix*, 51: 43–75.

Perkins, John. 2010. "Chemistry Courses, the Parisian Chemical World and the Chemical Revolution." *Ambix*, 57: 27–47.

Perkins, John. 2013. "Sites of Chemistry in the Eighteenth Century." *Ambix*, 60: 95–98.

Petterschmidt, Luc. 2010. "Théologie naturelle et philosophie chimique dans la *Siris* de Berkeley." *Dix-huitième siècle*, 42: 417–32.

Pluche, Noel Antoine. 1749. *Le spectacle de la nature, ou entretiens sur les particularités de l'histoire naturelle*, vol. 1. Paris: Veuve Estienne.

Poirier, Jean-Pierre. 1996. *Lavoisier: Chemist, Biologist, Economist*. Philadelphia, PA: University of Pennsylvania Press.

Porter, Roy. 1980. "The Terraqueous Globe." In George S. Rousseau and Roy S. Porter (eds), *The Ferment of Knowledge*. Cambridge: Cambridge University Press.

Powers, John C. 2012. *Inventing Chemistry: Herman Boerhaave and the Reform of the Chemical Arts*. Chicago, IL: University of Chicago Press.

Powers, John C. 2014. "Measuring Fire: Herman Boerhaave and the Introduction of Thermometry into Chemistry." *Osiris*, 29: 158–77.

Powers, John C. 2015. "Leiden Chemistry in Edinburgh: Herman Boerhaave, James Crawford, and Andrew Plummer." In Robert G.W. Anderson (ed.), *Cradle of Chemistry: The Early Years of Chemistry at the University of Edinburgh*. Edinburgh: Birlinn.

Priestley, Joseph. 1767. *History and Present State of Electricity*. London: Johnson.

Priestley, Joseph. 1769. *A New Chart of History*. London: Johnson.

Priestley, Joseph. 1774. *Experiments and Observations on Different Kinds of Air*. London: Johnson.

Priestley, Joseph. 1775. *Philosophical Empiricism: Containing Remarks on a Charge of Plagiarism*. London.

Priestley Joseph. 1775–1777. *Experiments and Observations on Different Kinds of Air*. 2nd ed., 3 vols. London: Johnson.

Priestley, Joseph. 1777. *Disquisitions Relating to Matter and Spirit*. London: Johnson.

Priestley, Joseph. 1785. *Reflections on the Present State of Free Inquiry in this Country*. London: Johnson.

Priestley, Joseph. 1790. *Experiments and Observations on Different Kinds of Air, and other Branches of Natural Philosophy: In Three Volumes*. Birmingham: Pearson and Johnson.

Principe, Lawrence M. 2001. "Wilhelm Homberg: Chymical Corpuscularism and Chrysopoeia in the Early Eighteenth Century." In Christoph Lüthy, John E. Murdoch, and William R. Newman (eds), *Late Medieval and Early Modern Corpuscular Matter Theories*. Leiden: Brill.

Principe, Lawrence M. 2007. "A Revolution Nobody Noticed: Changes in Early Eighteenth-Century." In Lawrence M. Principe (ed.), *New Narratives in Eighteenth-Century Chemistry*. Dordrecht: Springer.

Principe, Lawrence M., and William R. Newman. 2001. "Some Problems with the Historiography of Alchemy." In William R. Newman and Antony Grafton (eds), *Secrets of Nature: Astrology and Alchemy in Early Modern Europe*. Cambridge, MA: MIT Press.

Ragland, Evan. 2008. "Experimenting with Chymical Bodies: Reinier de Graaf's Investigations of the Pancreas." *Early Science and Medicine*, 13: 615–64.

Ragland, Evan. 2017. "Experimental Chemical Medicine and Drug Action in Mid-Seventeenth-Century Leiden." *Bulletin for the History of Medicine*, 91: 331–61.

Rayner-Canham, Marelene F., and Geoffrey Rayner-Canham. 1998. *Women in Chemistry: Their Changing Roles from Alchemical Times to the Mid-Twentieth Century*. Philadelphia, PA: Chemical Heritage Foundation.

Réaumur, René-Antoine Ferchault de. 1716–1727 [1888]. *Réflexions sur l'utilité dont l'Académie des sciences pourroit être au Royaume, si le Royaume luy donnoit les Secours dont elle a besoin*. In Ernest Maindron, *L'Académie des sciences*. Paris: Alcan.

Rey, Anne-Louise, and Philippe Huneman. 2007. "La controverse Leibniz-Stahl dite Negotium otiosum." *Bulletin d'histoire et d'épistémologie des sciences de la vie*, 14: 213–38.

Riello, Giorgio. 2010. "Asian Knowledge and the Development of Calico Printing in Europe in the Seventeenth and Eighteenth Centuries." *Journal of Global History*, 5: 1–28.

Riskin, Jessica. 2002. *Science in the Age of Sensibility: The Sentimental Empiricists of the French Enlightenment*. Chicago, IL: University of Chicago Press.
Risse, Guenter B. 1992. "Medicine in the Age of Enlightenment." In Andrew Wear (ed.), *Medicine in Society: Historical Essays*. Cambridge: Cambridge University Press.
Roberts, Lissa. 1991a. "A Word and the World: The Significance of Naming the Calorimeter." *Isis*, 82: 198–222.
Roberts, Lissa. 1991b. "Setting the Table: The Disciplinary Development of Eighteenth-Century Chemistry as Read through the Changing Structure of Its Tables." In Peter Dear (ed.), *The Literary Structure of Scientific Argument*. Philadelphia, PA: University of Pennsylvania Press.
Roberts, Lissa. 1993. "The Death of the Sensuous Chemist: The New Chemistry and the Transformation of Sensuous Technology." *Studies in the History and Philosophy of Science*, 26: 503–29.
Roberts, Lissa. 1995. "The Death of the Sensuous Chemist: The 'New' Chemistry and the Transformation of Sensuous Technology." *Studies in History and Philosophy of Science*, 26: 503–29.
Roberts, Lissa. 2008. "Chemistry on Stage: G.F. Rouelle and the Theatricality of Eighteenth-Century Chemistry." In Bernadette Bensaude-Vincent and Christine Blondel (eds), *Science and Spectacle in the European Enlightenment*. Aldershot: Ashgate.
Roberts, Lissa. 2014. "Practicing Oeconomy in the Second Half of the Long Eighteenth Century." *History and Technology*, 30: 133–48.
Roberts, Lissa, and Joppe van Driel. 2017. "The Case of Coal." In Lissa Roberts and Simon Werrett (eds), *Compound Histories: Materials, Governance and Production, 1760–1840*. Leiden: Brill.
Roberts, Lissa, and Simon Werrett (eds). 2017. *Compound Histories: Materials, Governance and Production, 1760–1840*. Leiden: Brill.
Robinson, Eric, and Douglas McKie (eds). 1970. *Partners in Science: Letters of James Watt and Joseph Black*. Cambridge, MA: Harvard University Press.
Robison, John (ed.). 1803. *Lectures on the Elements of Chemistry, by Joseph Black, M.D.* 2. vols. Edinburgh.
Roche, Daniel. 1988. *Les Républicains des lettres. Gens de culture et lumières au XVIIIe siècle*. Paris: Fayard.
Rosner, Lisa. 1991. *Medical Education in the Age of Improvement*. Edinburgh: Edinburgh University Press.
Rousseau, Jean-Jacques. [1782] 1995. *The Confessions and Correspondence*. Edited by Christopher Kelly, Roger D. Masters, and Peter G. Stillman. In *Collected Writings of Rousseau*, vol. 5. Boston, MA: University Press of New England.
Rousseau, Jean-Jacques. 1999. *Les institutions chymiques*. Paris: Fayard.
Rowe, David J. 1983. *Lead Manufacturing in Britain: A History*. London: Croom Helm.
Rudwick, Martin J. S. 2005. *Bursting the Limits of Time: The Reconstruction of Geohistory in the Age of Revolution*. Chicago, IL: University of Chicago Press.
Schabas, Margaret. 2005. *The Natural Origins of Economics*. Chicago, IL: University of Chicago Press.
Schaffer, Simon. 1990. "Measuring Virtue: Eudiometry, Enlightenment and Pneumatic Medicine." In Andrew Cunningham and Roger French (eds), *The Medical Enlightenment of the Eighteenth Century*. Cambridge: Cambridge University Press.

Schaffer, Simon, and Larry Stewart. 2005. "Vigani and After: Chemical Enterprise in Cambridge, 1680–1780." In Mary Archer and Christopher Haley (eds), *The 1702 Chair of Chemistry at Cambridge: Transformation and Change*. Cambridge: Cambridge University Press.

Schiebinger, Londa L. 2004. *Plants and Empire: Colonial Bioprospecting in the Atlantic World*. Cambridge, MA: Harvard University Press.

Schleiff, Hartmut and Peter Konečný (eds). 2013. *Staat, Bergbau und Bergakademie: Montanexperten im 18. und frühen 19. Jahrhundert*. Stuttgart: Franz Steiner.

Schofield, Robert E. 1997. *The Enlightenment of Joseph Priestley: A Study of his Life and Work from 1733 to 1773*. University Park, PA: Pennsylvania State University Press.

Schofield, Robert E. 2004. *The Enlightened Joseph Priestley: A Study of his Life and Work from 1773 to 1804*. University Park, PA: Pennsylvania State University Press.

Schranz, Kristen M. 2014. "The Tipton Chemical Works of Mr James Keir: Networks of Conversants, Chemicals, Canals and Coal Mines." *International Journal for the History of Engineering & Technology*, 84: 248–73.

Schrøder, Michael. 1969. *The Argand Burner: Its Origin and Development in France and England 1780–1800*. Odense: Odense University Press.

Serrano, Elena. 2013. "Chemistry in the City: The Scientific Role of Female Societies in late Eighteenth-Century Madrid." *Ambix*, 60: 139–59.

Serrano, Elena. 2017. "Spreading the Revolution: Guyton's Fumigating Machine in Spain: Politics, Technology, and Material Culture (1796–1808)." In Lissa Roberts and Simon Werrett (eds), *Compound Histories: Materials, Governance and Production, 1760–1840*. Leiden: Brill.

Shapin, Stephen, and Simon Schaffer. 1985. *Leviathan and the Air-Pump: Hobbes, Boyle, and the Experimental Life*. Princeton, NJ: Princeton University Press.

Shaw, Peter (ed.). 1727. A *New Method of Chemistry, including the Theory and Practice of the Art, a Translation of Boerhaave's Institutiones Chemiæ*. London.

Shaw, Peter. 1734. *Chemical Lectures, Publickly Read at London in the Years 1731 and 1732, and Since at Scarborough in 1733*. London: Shuckburgh and Osborne.

Shaw, Peter, and Francis Hauksbee. 1731. *An Essay for Introducing a Portable Laboratory*. London: Osborn and Longman.

Sheller, Mimi. 2003. *Consuming the Caribbean: From Arawaks to Zombies*. London: Routledge.

Siegfried, Robert. 2002. *From Elements to Atoms: A History of Chemical Composition*. Philadelphia, PA: American Philosophical Society.

Siegfried, Robert, and Betty Jo Dobbs. 1968. "Composition, a Neglected Aspect of the Chemical Revolution." *Annals of Science*, 24: 275–93.

Simon, Jonathan. 2002. "Authority and Authorship in the Method of Chemical Nomenclature." *Ambix*, 49: 207–27.

Simon, Jonathan. 2005. *Chemistry, Pharmacy and Revolution in France, 1777–1809*. London: Routledge.

Simon, Jonathan. 2014. "Pharmacy and Chemistry in the Eighteenth Century: What Lessons for the History of Science?" *Osiris*, 29: 283–97.

Siskin, Clifford, and William Warner (eds). 2010. *This is Enlightenment*. Chicago, IL: University of Chicago Press.

Smeaton, William S. 1966. "The Portable Chemical Laboratories of Guyton de Morveau, Cronstedt and Göttling." *Ambix*, 13: 84–91.

Smith, Adam. 1795. *Essays on Philosophical Subjects*. London: Cadell.

Smith, Cyril Stanley. 1964. "The Discovery of Carbon in Steel." *Technology and Culture*, 5: 149–75.
Smith, John Graham. 1979. *The Origins and Early Development of the Heavy Chemical Industry in France*. Oxford: Clarendon Press.
Smollett, Tobias. 1769. *The History and Adventures of an Atom*. London: Cook.
Spary, Emma C. 2012. *Eating the Enlightenment: Food and the Sciences in Paris, 1670–1760*. Chicago, IL: University of Chicago Press.
Spary, Emma C. 2014. *Feeding France: New Sciences of Food, 1760–1815*. Cambridge: Cambridge University Press.
Spector, Céline. 2013. "Essay on Taste. Essai sur le goût." In *A Montesquieu Dictionary*. Available online: http://dictionnaire-montesquieu.ens-lyon.fr/en/the-dictionary/.
Stahl, Georg Ernst. 1718. *Zufällige Gedancken und nützliche Bedenken über den Streit von dem so-genannten Sulphure*. Halle: Waysenhaus.
Stahl, Georg Ernst. 1720. *Einleitung zur Grund-Mixtion derer unterirdischen mineralischen und metallischen Cörper*. Leipzig: Eysseln.
Stahl, Georg Ernst. 1730. *Philosophical Principles of Universal Chemistry: Or, the Foundation of a Scientific Manner of Inquiry into Preparing the Natural and Artificial Bodies for the Uses of Life*. London: Osborne and Longman.
Stauffer, Robert C. 1960. "Ecology in the Long Manuscript Version of Darwin's *Origin of Species* and Linnaeus' *Oeconomy of Nature*." *Proceedings of the American Philosophical Society*, 104: 235–41.
Stewart, Larry. 1992. *The Rise of Public Science*. Cambridge: Cambridge University Press.
Stock, John T. 1969. *Development of the Chemical Balance*. London: HMSO.
Stroup, Alice. 1979. "Wilhelm Homberg and the Search for the Constituents of Plants at the 17th-Century Académie Royale des Sciences." *Ambix*, 26: 184–201.
Stroup, Alice. 1990. *A Company of Scientists: Botany, Patronage, and Community at the Seventeenth-Century Parisian Royal Academy of Sciences*. Berkeley, CA: University of California Press.
Strasser, Susan. 2014. *Waste and Want: A Social History of Trash*. New York, NY: Henry Holt.
Szabadváry, Ferenc. 1966. *History of Analytical Chemistry*. Oxford: Pergamon.
Szalay, Gabriella. 2019. "Paper Trials, Multiple Masculinities, and the Economy of Honor." In Clara Bittel, Elaine Leong, and Christine von Oertzen (eds), *Working with Paper: Gendered Practices in the History of Knowledge*. Pittsburgh, PA: University of Pittsburgh Press.
Taylor, Georgette. 2008. "Making Out a Common Disciplinary Ground: The Role of Chemical Pedagogy in Establishing the Doctrine of Affinity at the Heart of British Chemistry." *Annals of Science*, 65: 465–86.
Thackray, Arnold. 1970. *Atoms and Powers: An Essay on Newtonian Matter-Theory and the Development of Chemistry*. Cambridge, MA: Harvard University Press.
Thébaud-Sorger, Marie. 2009. *L'aérostation au temps des Lumières*. Rennes: Presses Universitaires de Rennes.
Thébaud-Sorger, Marie. 2018. "Capturing the Invisible: Heat, Steam and Gases in France and Great Britain, 1750–1800." In Lissa Roberts and Simon Werrett (eds), *Compound Histories: Materials, Governance and Production, 1760–1840*. Leiden: Brill.
Thompson, Helen. 2017. *Fictional Matter: Empiricism, Corpuscles, and the Novel*. Philadelphia, PA: University of Pennsylvania Press.

Todericu, Doru. 1984. "Balthazar Georges Sage (1740–1824): Chimiste et minéralogiste français, fondateur de la première Ecole des Mines (1783)." *Revue d'histoire des sciences*, 37: 29–46.

Tomory, Leslie. 2009. "The Origins of Gaslight Technology in Eighteenth-Century Pneumatic Chemistry." *Annals of Science*, 66: 473–96.

Tomory, Leslie. 2012. *Progressive Enlightenment: The Origins of the Gaslight Industry, 1780–1820*. Cambridge, MA: MIT Press.

Tomory, Leslie. 2017. *The History of the London Water Industry, 1580–1820*. Baltimore, MD: Johns Hopkins University Press.

Tribe, Keith. 2005. "Oeconomic History: An Essay Review." *Studies in the History and Philosophy of Science*, 36: 586–97.

Tylecote, Ronald F. 1992. *A History of Metallurgy*. London: Institute of Metals.

Uglow, Jenny. 2011. *The Lunar Men: The Inventors of the Modern World 1730–1810*. London: Faber and Faber.

Van den Boogert, Maurits H. 2010. *Aleppo Observed: Ottoman Syria through the Eyes of Two Scottish Doctors*. Geneva: The Arcadian Library.

Vankin, Deborah. 2018. "'Blue Boy' Revisited: The Huntington is Saving its 18th-century Masterpiece – and You Get to Watch." *Los Angeles Times*, September 14, 2018. Available online: https://www.latimes.com/entertainment/arts/la-ca-cm-project-blue-boy-20180914-story.html.

Van Spronsen, Jan W. 1975. "The Beginning of Chemistry." In T.H. Lunsingh Scheurleer and G.H.M. Posthumus Meyjes (eds), *Leiden University in the Seventeenth Century: An Exchange of Learning*. Leiden: Brill.

Venel, Gabriel François. 1753. "Chymie ou chimie." In Denis Diderot and Jean le Rond d'Alembert (eds), *Encyclopédie, ou dictionnaire raisonné des sciences, des arts et des métiers*, vol. III, 408–37. Paris: Briasson.

Vidal, Mary. 1995. "Lavoisier Among the Moderns: Art, Science and the Lavoisiers." *Journal of the History of Ideas*, 56: 595–623.

Vogel, Jakob. 2010. "Locality and Circulation in the Habsburg Empire: Disputing the Carlsbad Medical Salt, 1763–1784." *British Journal for the History of Science*, 43: 589–606.

Vogel, Virgil J. 2013. *American Indian Medicine*. Norman, OK: University of Oklahoma Press.

Wakefield, Andre. 2010. "Police Chemistry." *Science in Context*, 12: 231–67.

Weeks, Mary Elvira. 1933. *The Discovery of the Elements*. Easton, PA: Journal of Chemical Education.

Werrett, Simon. 2010. *Fireworks: Pyrotechnic Arts and Sciences in European History*. Chicago, IL: University of Chicago Press.

Werrett, Simon. 2013. "Green is the Colour: St. Petersburg's Chemical Laboratories and Competing Visions of Chemistry in the Eighteenth Century." *Ambix*, 60: 122–38.

Werrett, Simon. 2017. "Household Oeconomy and Chemical Inquiry." In Lissa Roberts and Simon Werrett (eds), *Compound Histories: Materials, Governance and Production, 1760–1840*, 35–56. Leiden: Brill.

Werrett, Simon. 2019. *Thrifty Science: Making the Most of Materials in the History of Experiment*. Chicago, IL: University of Chicago Press.

Willey, Basil. 1940. *The Eighteenth Century Background: Studies on the Idea of Nature in the Thought of the Period*. London: Chatto and Windus.

Willies, Lynn. 1991. "Lead: Ore Preparation and Smelting." In Ronald F. Tylecote and Joan Day (eds), *The Industrial Revolution in Metals*. London: Institute of Metals.

Wills, Hannah. 2019. "The Diary of Charles Blagden: Information Management and the Gentleman of Science in Eighteenth-Century Britain." Ph.D. Thesis, University College London.

Wilson, George. 1709. *A Compleat Course of Chemistry, Containing not only the Best Chymical Medicines, but also Great Variety of Useful Observations*. 3rd ed. London: John Bayley.

Wilson, George. 1851. *The Life of the Hon'ble Henry Cavendish*. London: Cavendish Society.

Wilson, Wendell E. 1994. *The History of Mineral Collecting 1530–1799*. Tucson, AZ: The Mineralogical Record.

Wisniak, Jaime. 2005. "Matches: The Manufacture of Fire." *Indian Journal of Chemical Technology*, 12: 369–80.

Young, Arthur. 1792. *Travels during the years 1787, 1788 and 1789*. London.

LIST OF CONTRIBUTORS

Bernadette Bensaude-Vincent is Professeur Emerita at the Université Paris 1 – Panthéon Sorbonne, France.

Marco Beretta is Professor of History of Science at the Università di Bologna, Italy.

Victor D. Boantza is Associate Professor in the Program in History of Science and Technology, University of Minnesota, Minneapolis, MN, USA.

John R.R. Christie is on the Faculty of History, University of Oxford, UK.

Matthew Daniel Eddy is Professor in the Department of Philosophy, Durham University, UK.

Ursula Klein is Professor and senior researcher at the Max Planck Institute for the History of Science, Berlin, Germany.

John C. Powers is Associate Professor in the Department of History, Virginia Commonwealth University, Richmond, VA, USA.

Leslie Tomory is Professor of Practice in the Department of History, McGill University, Montreal, Canada.

INDEX

Aberdeen University 115
Åbo University 166
academic status 78, 94, 101, 103
acid–alkali reactions 57
acidity 43
affinity tables 12–13, 16, 32–3, 43, 58, 107, 172–3, 198
Afzelius, Johan 86–7
agriculture 15, 21, 94, 103, 115, 118, 125, 152–3, 202
air, different kinds of 9, 34, 63–4, 67, 83–4, 134–5
alchemy 1, 98, 179, 202
  persistence of 183–9
d'Alembert, Jean 10, 48, 76, 103, 142, 144
alkalis 6, 153–5
alum 7
analytical methods 6, 33–5, 38
Anderson, Robert 170
antimony 6
apothecaries 74, 106
  teaching of and by 158–64, 167–9
apparatus, chemical 10–11, 45, 48–52, 62–9, 71–91, 116, 122, 182–3, 186
  portable 88
applied chemistry 5
apprenticeship in chemistry 158–62
  limitations of 159–60
Argand, Aimé 152
Aristotle and Aristotelianism 24, 28, 37, 129
artisans 100–2, 114, 137–8, 159

Astbury, John 150
atomism 32
attrition 48
audiences for chemistry 95–6, 167–71

Bacon, Francis 99, 103, 170
Bacon, Roger 98
Baptist's Head Coffee House 113–14, 135
Barbauld, Anna Letitia 176, 193, 196–7
Barchusen, Johann Conrad 164
Baumé, Antoine 78–81, 106, 161, 172
Beccari, Jacopo Bartolomeo 153
Becher, Johann Joachim 25–9, 37
Beckman, Johan 102
Beddoes, Thomas 116, 193
Beguin, Jean 160, 167
Belgium 105
Berch, Anders 86
Beretta, Marco 186, 188
Bergman, Torbern 8, 59–60, 84–8, 120, 131, 149, 200–1
Berkeley, George 108–9
Bernardi, Bruno 110
Berthelot, Marcellin 46, 48, 69
Berthollet, Claude-Louis 13, 65, 107, 111, 140–2, 149, 201
Berzelius, Jacob 60
Bianchi, Jacques 117
bismuth 6
Black, Joseph 9, 63–7, 73, 82, 94, 105–6, 120, 124, 132, 155, 161, 165–6, 173, 176–9

Blackstone, Sir William 194
Blagden, Sir Charles 118
Blancard, Stephan 118
bleaching 139–41
blood circulation 129
blowpipes 60, 85–6
Boerhaave, Herman 55, 57, 73, 82, 98, 109–10, 121, 164–6, 169–72, 179
Bohn, Johannes 165
books *see* literary works
Boscovich, Ruggiero 108, 111
Böttger, Johann Friedrich 123, 150
Boulduc, Simon 53–6
Boulton, Matthew 116–17, 149
Bourdelin, Claude 52–4
Boyle, Robert 4, 32, 52, 57, 61–2, 66, 99, 117, 172, 180
Brandt, Hennig 152, 184
Brant, Georg 169
brass 6, 146
bronze 6
Brougham, Henry 176–8
Bruce, Jacob 116
Bucquet, Jean Baptiste Michel 173
Buffon, Comte de 107
Burchusen, Conrad 182–3
Burke, Edmund 126–7, 190
burning lenses 84, 168

calcareous earths 7
calcination 49
caloric 41–4
calx 13, 29, 36, 39, 66–7, 71, 171
cameralism 15, 102–3, 125
candles 151–2
carbon dioxide 9, 42, 58, 63, 82, 85, 123, 134, 176
caricatures 190, 193
cast iron 148–9
Cavendish, Henry 9, 64–8, 83, 197
Chalmers, Alan 33
Champion, William 146–7
Chang, Ku-Ming 25
Chang, Hasok 35
Chapral, Jean-Antoine 101, 140, 180
charcoal 7, 10, 13, 36–7, 43, 60, 67, 77, 79, 146–9, 154–5
Châtelet, Emilie 18
chemical arts 4, 86, 101–2

chemical change 11–12, 25, 67
chemical cultures 112
chemical industries 4, 7, 20, 137
  trends in 156
"chemical revolution" 13–14, 23, 40, 45–6, 52, 65
chemical substances 5–8, 11–12, 118, 121
  knowledge of 5
  processed 7–8
chemico-theology 127
chemistry
  behind the facade 105–8
  as a branch of natural philosophy 104
  as a conceptual–practical hybrid 46, 74, 78
  everyday form of 20
  expansion and reform of 45, 135, 158, 202
  golden age of 17, 99
  as an independent academic discipline 1–4, 70, 94
  origins of 98, 179
  of plants and animals 8
  popularity of 74, 95, 97
  pure and applied 5, 101
  rapid and dramatic changes in 112, 160–1
  sub-disciplines of 5, 8
chemists
  creative powers of 69
  with different backgrounds 3, 66
  employment of 114–15, 135
  image of 93–4, 97
  recognition of 3
  seen as *artistes* 102
Christie, John 98, 158
chymistry 1–2
cinchona bark/trees 21
Cisternay du Fay, Charles François de 143
cities 19, 21, 25, 102
climate 2, 135, 153
coagulation 49
coal 21, 27, 117, 133, 144–52
Cochin, Charles-Nicolas the Younger 118
Cochrane, Archibald 153
coction and *decoction* 50
cohobation 50
collature 49
colonialism/colonies 124, 138

combustion, theory of 66–7
commercial success 21
communities of chemists 3–4, 174
"compositionism" 35
compounds, chemical 11–14, 24, 28–31
Condillac, Etienne Bonnot de 108
Congreve, William 155
consumer goods 156
consumerism 121–4
Cook, James 124
Cooksworthy, William 150
Cooper, Thomas 193
copper 145–6
corpuscular theory 4, 11, 25–6, 32, 105
corrosion 36
courts, princely 1
craft practice 4, 158
Crell, Lorenz 3, 16, 120, 160
Croll, Oswald 121
Cromwell, Oliver 191
Cronstedt, F. 8, 85
crop rotation 133
Crosland, Maurice 40
Cruickshank, Andrew 191
Cullen, William 94, 99, 105, 107, 161, 165–6, 173, 179–81
cultural ascent of chemistry 97, 100
Cyprianus, Abraham 163

Dalton, John 112
Dandolo, Vicenzo 101, 105
Darby, Abraham 146, 148
Darwin, Erasmus 184
David, Jacques-Louis 176, 185–9
Davy, Humphry 69–70, 83, 108
decantation 48–9
decomposition 11–12, 27–9
deflagration 49
de Luc, Jean André 131
demonstrations, chemical 68, 74, 96, 168, 172, 178
dentition 196
dephlegmation 49
Desaguliers, John Theophilus 117
detonation 49
Devonshire, Duchess of 116, 118
diagrams 16, 19, 172–3
dictionaries, chemically-oriented 118–20
Diderot, Denis 10, 48, 76–9, 96, 102, 109, 142, 144, 168

Diesbach, Johann Jacob 141
Dillinger, Bergrath 147
Diluvialism 131
distillation 46, 50–4, 57
Dodart, Denis 53
domestic goods 150–3
Dossie, Robert 72, 181
"dry" operations 49
Duclos, Samuel 4, 52–4
Duhamel du Monceau, Henri-Louis de 154
Duncan, Alistair 32
Dupin, Claude and Louise Fontaine 97
dyeing 141–3

earths 7, 9
East India Company, British 122, 155
East India Company, Dutch 122, 155
ebullition 49
ecology 131–5
    divine 132
Edinburgh 170
    Incorporation of Surgeons 161
    Royal College of Physicians 124
    University of 15, 73, 82, 94, 99, 116, 132, 161, 166
education of children 117
effluvium thesis 129–31
Ekaterina Romanova Dashkova, Princess 117
"elastick" and "unelastick" air-particles 63
elective affinity 198–9
electricity 38, 40, 83–4
electrum 6
elements
    Aristotelian (*earth, water, air* and *fire*) 24, 46, 109–10
    chemical 11, 14, 26, 33, 35
Elhuyar, Juan José de 85
elixation 50
elutriation 49
empires see colonialism/colonies
the *Encyclopédie* 16, 76–7, 103–7, 112, 142, 144, 181
engravings 77, 118
Enlightenment movement 14–17, 94, 97, 112
entrepreneurs 20, 73, 82, 105, 114–16, 121, 133
environment, the 128–31, 135
epistemology 110

equipment *see* apparatus
eroticization 188
eudiometry 83, 100–1
experiential knowledge 53, 68
experimentation 45–50, 65–9, 76–82, 88–91, 95, 97, 102, 114, 122
    cautious approach to 79
    difference from theory 11
expertise, hybridity of 114
explosions 125–6

factory system 20, 142, 156
farming practice 152–3
Ferguson, Adam 108
fertilizers 153
Fielding, Henry 121
filtration 49
fire 50, 53–4
Fishwick, Richard 148
fixed air 9, 63–4, 67, 82, 123–4, 134, 176, 178
fixed alkali 5
Fontanelle, Bernard 98, 104, 109–10, 168
Fortin, Jean Nicolas 66
Fourcroy, Antoine François 9, 13, 65, 91, 101, 128, 173
Fox, Charles James 191
France 2–3, 7–8, 19, 46, 75, 105, 109, 115, 125, 128, 140–1, 153–4, 160
    Académie Royale des Sciences 30, 69, 73, 80, 94, 98, 101, 106, 120, 154, 161–2, 168, 172
    Jardin des Plantes 2, 160
    Jardin du Roi 56, 75, 96, 110, 167–9
Franklin, Benjamin 195
fraud 98
Frederick I of Prussia 162–3
Frederick II ("the Great") 7
French Revolution 190–1, 202
fulmination 49
fumes, toxic 78
furnaces 3, 10, 21, 45–6, 71–2, 76–9, 81, 86, 116, 122, 129, 145–9, 159, 181

Gadolin, Johan 87
Gahn, Johan Gottlob 87
Gainsborough, Thomas 20
Galen 179
Garbett, Samuel 140
gases 9, 60–1

Gaubius, Hieronymus David 163
gender 24, 189, 197–200
Geoffroy, Claude Joseph 53, 55
Geoffroy, Etienne-François 12, 31–3, 53, 58–9, 98, 162, 172–3
geology 21, 94, 128–31, 166
George III, King 126
Germany 3, 9, 15, 60, 73, 110–11, 122, 125, 159–63, 166, 169
Gillray, James 19, 126–7, 190–1
girls 17–19, 117–18
glass-making 150–1
Glauber, Rudolf 77, 152
Godfrey, Ambrose 181–2
Goethe, Johann Wolfgang von 18, 121, 176, 198–202
Golinski, Jan 158, 170
Gosse, Henri-Albert 165
Gothic revivalism 184, 202
Göttingen University 102
government employment 115–16
gravimetric methods 64, 68–9, 107–8
Grotius, Hugo 194–5
Guerlac, Henry 166
Guettard, Jean-Etienne 88, 166
guilds 72, 81
gunpowder 155, 191

Hales, Stephen 9, 34, 61–4, 67, 77, 84, 152
Halle University 165
Harvey, William 129
Hauksbee, Frances 72
Haüy, René-Just 8–9
heat and heating 7–13, 21, 26, 34–6, 38–43, 49–55, 59–69, 74–84, 104–7, 112, 122–3, 129, 131, 142–50, 155, 182–3
Heberden, William 133
Hegel, Georg Wilhelm Friedrich 111
Heinitz, Friedrich Anton von 9
Hellot, Jean 142
Hermbstädt, Siegmund Friedrich 133
Hermbstädt, Sigismund 162
Hippocrates 128
history and historians of chemistry 13, 40, 43, 71, 157, 179, 184
Hjelm, Peter Jacob 84–7
Hobbes, Thomas 110, 125
Hoffmann, Friedrich 165

Hoffmann, Gottfried 137
Hogarth, William 115, 123
Hohenheim, Theophrastus von 46
Holker, John 140
Holmes, Frederic L. 13, 31, 157
Homberg, Wilhelm 2, 26, 30, 53–4, 57, 98, 109
Home, Henry 153
Hope, Thomas Charles 82
Hopkinson, Francis 117
household activities 20, 118
household goods 122
Hufbauer, Karl 3, 159, 162, 166
Hume, David 110
Hume, Francis 140
humid analysis 131
Huntsman, Benjamin 149
Hutton, James 129–31
hydrology 28, 128–31
hypnotism 19

imperialism 21
improvement of society 95, 102–3, 125, 202
industrial revolution 20–1, 73, 116, 133, 138
infusion 49
instruments 91 *see also* apparatus
international trade 156
ipecacuanha plant 54
iron manufacturing 148–9
Italy 105

Jena University 165
Johnson, Joseph 196
journals 3, 16, 91, 120
Juvenal 194

Kant, Immanuel 110–11
kaolin 122–3
Kay, John 176–8
Keill, John 131
Keir, Amelie 19
Keir, James 155, 184, 193
Kerr, Robert 178
Kirwan, Richard 43
Klaproth, Martin 9, 18, 162
Klein, Ursula (co-editor) 162

Knight, David 111
Krünitz, Johann Georg 10
Kunckel, Johann 28

laboratories 3–4, 9–11, 45, 48, 71–81, 84–8, 91, 97
   definition of 10
   private 164–5
   teaching in 174
   ventilation of 78
   see also representations
la Brosse, Guy de 167
Laing, William 114–15
laissez-faire economics 125
Landriani, Marsilio 83
Laplace, Pierre-Simon 65, 88, 111
La Révelière-Lepaux, Louis Marie de 126
Latin language 120
Lavoisier, Antoine-Laurent 6, 9–14, 25, 39–47, 52–3, 56–60, 63–9, 75, 87–91, 93, 99–103, 106–8, 111–12, 116, 127–8, 134–5, 155, 157–8, 166, 168, 173, 176–80, 185–90
   concept of elements and principles 34–6
   innovations by 14
   theory of combustion and calcination 42–4
Lavoisier, Marie-Anne Paulze 18, 66, 90–1, 118, 176, 183–9, 197
lead (metallic element) 147–8
learned societies 16
Leblanc, Nicolas 151, 154–5
Lebon, Philippe 152
Le Bouyer de Fontenelle, Louis-Bernard 95
Leclerc, Georges-Louis 107
lectures 18, 101, 158, 167–70, 174
Leibniz, Gottfried Wilhelm 108–10
Leiden University 164–6, 169, 172, 179
Leipzig University 165
Lémery, Nicolas 3, 30, 46, 50, 53, 56–7, 151, 160–1, 167–8, 171
Le Mort, Jacob 164
Le Sage, René 195–6
lessons in chemistry, private 165
letters 16, 87, 117, 120, 129, 189, 191, 198
Levant Company 122
Lewis, William 181–3

INDEX 235

lexicons 118–20
Leyden Phial, the 195
Libavius, Andreas 2, 167
lime 153
limewater 63, 79
limestone 7, 63–4, 131, 154
Lindsey, Theophilus 191
Linnaeus, Carl 15–16, 84–5, 108, 114
literature 17–18, 121, 176, 193–202
lixiviation 49
local knowledge 4–5
local traditions 73
Locke, John 108
London 133
    theatres in 20
Lucretius 125
Lund University 166
Lyon, John 129

maceration 49
Macintosh, Charles 141, 144
Macquer, Pierre-Joseph 33–5, 38–43, 46, 53, 57, 78–80, 88, 93–101, 107, 118, 142, 150, 173, 180
Madrid 118
Maets, Caerl de 164
Magellan, Jean-Hyacinthe de 106
magnetism 19, 95, 104, 106, 112, 127
Maier, Michael 98
Malouin, Paul-Jacques 147
manufacturing 20, 181
Marcet, Jane 19
Marggraf, Andreas 55, 59–60
Marggraf, Sigismund 161
matches (made from phosphorus) 152
material culture of chemistry 103
matter, theory of 23–5, 28–9, 33, 128, 172
Mayow, John 62
mechanics 107
mechanization 138, 142–3
medicine, study and practice of 1, 55, 63, 72–3, 94, 121, 128, 163–5
Megnié, Pierre 66, 89–90
Meinel, Christoph 164
Mercier, Sébastien 95
mercury 46, 77, 123
Mesmer, Franz Anton 19

metals 6–8, 11–14, 26–7, 29, 30, 36–7, 39, 40, 42–3, 46, 49, 53, 57, 58, 71, 78–9, 86, 110, 116, 122, 124, 129, 145–6, 171
metallurgy 145
metalworking 4
metaphor 95, 125, 127
meteorology 21, 84, 104, 128, 134
Metz 169
Metzger, Hélène 25
Meusnier, Jean Baptiste 68, 89, 101
miasmas 133
minerals 8, 9, 15, 17, 26–7, 29, 37, 46, 56, 60, 73, 85–8, 102, 114, 123–4, 139, 141, 132, 166, 171
mineral acids 7, 30, 63
mineral chemistry 56–60
mineral collections 85–6
mineral waters 4, 8, 21, 85, 88, 105
mineralogy 4, 8, 69, 82, 85, 94, 101, 104–5, 116, 129, 162–3, 169
    chemical 8
mining 8, 20
mixts 24, 27–31, 54, 104
monarchs as patrons 116
Monge, Gaspard 149
Montesquieu, Charles-Louis de Secondat 95
mordants 141–5
Morveau, Guyton de 13, 65, 67, 100, 106–7, 128, 176
mosaic image 112
Moyes, Henry 18, 117
Murdoch, William 152
Muspratt, James 155

Nancy 169
natron 6
natural disasters 131
natural history 128–9, 166
natural philosophy 4, 25, 82, 84, 94–7, 103–4, 108, 111, 171
natural scientists 105
naturally-occurring substances 11, 24–5, 28–37
Naturphilosophie 18, 111
Necker de Saussure, Albertine 18
Neptunists 131

Netherlands, the 105
Neumann, Caspar 160–3, 169
"new chemistry" 65, 99–100, 186
Newton, Sir Isaac (and Newtonian science) 13, 32–3, 99, 104–11, 127, 180, 193, 198
Nicholson, William 16
nickel 8
"noble" metal 30
nomenclature 14, 120, 157, 186
novels 17–18, 23–5, 121, 176, 198–202

oeconomy 103
oil for lamps 151–2
Olmedo, Vincente 124
"organic" substances 56
Ovid 194
oxygen 35, 40–1, 53, 67, 84, 90, 106, 132, 157, 190

Paine, Tom 191
paintings 182, 184
Pallas, Simon Peter 124
Paracelsus 24–8, 37, 98, 179, 181
Paris 19, 74–5, 78, 96–8
Parker, William 84
Parmentier, Antoine-Augustin 153
patronage 114–17, 170
Pavel Michailovich Dashkov, Prince 117
pelicans 50–2
per ascensum and *per decensum* distillation 50
Percival, Thomas 132
periodicals 120, 133
Perkins, John 95, 100, 168
Peter the Great 116
petrogenesis 129
pharmacy 72, 74, 106, 160–3, 181
philosophers' stone 179, 183–4
philosophes 96
philosophical societies 113–14, 135
philosophy 108–12
    as an approach to chemistry 171–4
phlogiston theory 12–14, 27, 29, 34–44, 66–7, 99, 157, 172–3, 186, 190–3, 196–7
physics 48, 104, 106
Picardet, Claudine 18
pictic acid 141

pig iron 148–9
Pitt, William the Younger 190
plants 8, 15–17, 37–9, 49, 52–8, 102, 124, 134, 141, 152–6
plays 196
Pluche, Noel-Antoine 95, 109
Plummer, Andrew 166
pluralism 25
Plutonists 131
pneumatic chemistry 60–5, 77, 82–7, 97, 100, 105–7, 134
poetry 193–7
politics 125–8
pollution 132
porcelain 122, 150–1
portraiture 185–9
potatoes 153
Pott, Johann Heinrich 7, 123, 129–31, 160–1, 169
pottery 150
practical chemistry 4–6, 163–74
prestige of chemistry 72–3, 94, 108–12
Price, Richard 191, 193
Priestley, Joseph 9, 17, 40, 43, 53, 60–7, 75, 83–6, 97, 100–1, 106, 125–8, 132, 134, 176, 182–3, 190–7
Priestley, Mary 193
principles
    chemical 26–9, 34–5
    active and *passive* 46
"principlism" 35
printing and print culture 97, 120
problem-solving 21
professorships in chemistry 15, 72–3, 116, 163–6, 169
Prussia 160
    Academy of Sciences 73
Prussian blue dye 141
public image of chemistry 17–21, 106, 180
public lectures 2, 18, 97, 174, 181
purification 50

quicklime 7
Quinquet, Antoine Arnaud 165

Radcliffe, Ann 17
"rapports" 32
Ravenscroft, George 151
raw materials 7–9

Real Escula de Quimica 116
Réaumur, René de 73, 149, 151
rectification 50
regulation of chemical trades 72
Reid, Thomas 108
religion 17, 94, 191, 193, 196, 202
representations of chemistry 175–9
    literary 193–202
    showing laboratories 181–4, 190
Republic of Letters 120
retorts 51–2
revivification 49
revolutionary chemistry 65–70
Rinman, Sven 75–6
rivers 133
Robertson, William 117
Robison, John 99, 128, 173, 176–80
Roebuck, John 140, 155
Rolfinck, Werber 165
Rouelle, Guillaume-François 56, 75–9, 91, 96, 109–10, 166, 168, 172–3
Rousseau, Jean-Jacques 96–7, 109–10
Royal Institution, London 18–19, 69
Royal Society of London 16, 83, 101, 120, 193
Ruberg, Johann 147
Ruprecht, Anton von 169
Russell, Alexander Patrick 121
Russia 116

Sage, Balthazar-Georges 170
saltpetre 7, 155
salts 7, 12, 30–1, 43, 46, 57–8
Sartorius, C.F. 161–2
satire 190, 193
Sayers, James 191–3
Scarborough 170
Schäffer, Jacob Christian 16–17
Scheele, Carl 9, 87, 140, 152
Schelling, Friedrich 111–12
science
    authenticity and *independence* in 179
    chemistry regarded as 13, 18, 35, 42, 70, 72–6, 98–100, 180
    modern form of 98
    status of 180
    see also history and historians of chemistry
scientific revolution 40, 48, 176

Scotland 94, 105, 108, 114–15
secrecy 114, 122
Séguin, Armand 88–91
self-representation by chemists 178–81
Senebier, Jean 152
Sennert, Daniel 121
Sèvres porcelain factory 78
Shaw, Peter 72, 96, 98, 170, 172
Shelburne, Earl of 97, 191
Shelley, Mary 18
Siegfried, Robert 35
Simon, Jonathan 161
simple substances 14, 35–6, 41
smelting 6
Smith, Adam 180–1
smoke 133
Smollett, Tobias 121
soap 151
sociability 16, 117–21
social contract 110
social status 99–102, 114
soda 154–5
soil depletion 133
South America 85, 125
Spain 105, 116, 124
specific gravity 85
St. Petersburg 116
Stahl, Georg Ernst 4, 12–13, 25–9, 33–8, 103–4, 109–10, 117–18, 163–5, 170–2
"Stahlianism" 103–4
Stam, David 165
Stanhope, Lord 191, 193
Starkey, George 49–50, 66
state-sponsored projects 125
status of chemistry 3, 99, 115, 163, 165, 174; *see also* academic status
steel 149
Steno, Nicolas 131
Stewart, Dugald 108
"subtle" substances 6
supply networks 121
Swab, Anton von 85–6
Sweden 8, 72–3, 105, 166
synthesis 16, 33, 38, 57, 90, 154, 172, 198

tanning 7, 72, 100
teaching of chemistry 74–80, 117, 158–74
Tennant, Charles 141

textbooks, chemical 2–3, 16, 25, 30, 46, 98, 158, 164, 167–73, 178, 186
textile industries 20, 138–9
"theater of nature" metaphor 95
theoretical chemistry 11–13, 172, 174
Thirty Years War 145
tin and tinplate 147
tobacco 133
torrefaction 49
trade
    chemical synergies with particular industries 123
    linked to chemical industries 100, 137
    purity standards in 125
trade secrets 114, 122
transmutation of metals 11, 30, 32, 179
trituration 48
Trommsdorff, Bartholomäus 16
Trommsdorff, Johann 16, 162
Tschirnhaus, Ehrenfried Walther von 123
Turgot, Robert Jacques 96, 168
Turkey red dye 141

universities 1–2, 15, 94, 102–3, 110–11, 161–6, 169, 181
    integration of chemistry into the curriculum 163
Uppsala University 72, 165, 169
urban settings 18–21, 100, 103, 118, 132, 137
utilitarian science 101, 103
Utrecht University 164

van Helmont, Jan Baptist 61, 66
van Swieten, Gerard 165
Vandermonde, Alexandre-Théophile 149
Vauquelin, Nicolas Louis 8–9

Venel, Gabriel François 48, 101–2, 105–10, 168, 172, 181
Vesuvius 130
Vienna 19, 165
Vigani, Giovanni Francesco 164
visual culture 123, 176
vitriol 7
volcanoes and Vulcanists 129–31
Volta, Alessandro 88

Wallerius, Johan Gottschalk 101, 131, 153, 166, 169
Ward, Archer 148
Ward, Joshua 140
Warens, Madame de 96
water supply 132–3
Watson, Richard 191, 193
Watt, James 82, 116–17, 149, 155, 176
Wedgwood, Josiah 81–2, 150
weighing techniques 66, 69
Werner, Abraham Gottlob 131
Werrett, Simon 103
"wet operations" 30, 48–9
White, John 140
Wiegleb, Johann Christian 161–2
Willey, Basil 127
Wilson, George 171–2
wisdom, dignity of 104
women, roles and status of 18–19, 74–5, 96, 116–18
Wood, Charles and John 149
Woulfe, Peter 141
Wright, Joseph 183–4, 195
Wright, Thomas 18–19
wrought iron 148–9

Young, Arthur 89

zinc 146–7

Instruments (including apparatus, retort, tube, receiver, bottle, phial, glassware, pelican, alembic, scale, balance, barometers, thermometers, hygrometers, air pumps, eudiometers, and microscopes, etc.)

Meteorology (including measurement, scale, weights, grains, measures, barometers, thermometers, hygrometers, air pumps, eudiometers).